International Conference on Ice Class Vessels 2020

Online
4 November 2020

ISBN: 978-1-7138-8300-5

Printed from e-media with permission by:

Curran Associates, Inc.
57 Morehouse Lane
Red Hook, NY 12571

CURRAN ASSOCIATES INC. proceedings.com

Some format issues inherent in the e-media version may also appear in this print version.

The Royal Institution of Naval Architects
8-9 Northumberland Street
London, WC2N 5DA
United Kingdom

Phone: 020 7235 4622
Fax: 020 7259 5912

publications@rina.org.uk

Additional copies of this publication are available from:

Curran Associates, Inc.
57 Morehouse Lane
Red Hook, NY 12571 USA
Phone: 845-758-0400
Fax: 845-758-2633
Email: curran@proceedings.com
Web: www.proceedings.com

International Conference

Ice Class Vessels 2020

4th November 2020, Online Conference

PAPERS

THE ROYAL INSTITUTION OF NAVAL ARCHITECTS
8-9 Northumberland Street, London, WC2N 5DA, Telephone: +44 (0)20 7235 4622

The Royal Institution of
Naval Architects

INTERNATIONAL CONFERENCE

Ice Class Vessels 2020

4th November 2020, Online Conference

Papers published in these conference Proceedings are peer reviewed.

THE ROYAL INSTITUTION OF NAVAL ARCHITECTS
8-9 Northumberland Street
London WC2N 5DA

Telephone: 020 7235 4622
Fax: 020 7259 5912

CONTENTS

DESIGN AND CONSTRUCTION OF ICE CLASS VESSELS

S. Ghosh, Naval Architect and Marine Engineer, India

SUMMARY

Because of the anomalies of icy water, the philosophy of basic design and construction too assumes an intricate study of challenges. Perhaps it can be considered an extreme form of surface vessel design. Due to global warming and the resultant whittling of ice covers globally, the drifting ice is venturing into increased ocean traffic routes. So, this is a comprehensive paper with an eclectic range of aspects covered like climate change and environments, ice types, their properties, and behaviour; how these parameters in turn influence the design philosophy of these vessels. These crucial factors imbibed during detailed design and construction are also discussed. This is further continued into the domain of hull form design, materials, strengthening, failure considerations, critical points of maximum loading, vulnerable areas, etc. Also, though the underlying principles of design remain more or less the same, the degrees of these factors vary in the case of ice-going ships, ice-strengthened ships, and icebreakers, the most advanced form of an ice-capable ship, the nuances of which have also been briefly summarised. A brief note into the subject of Classification for Ice Ships is also included.

NOMENCLATURE

k_i = Thermal Conductivity
L_f = Latent Heat of fusion
P_i = Density of Ice
T_b = Temperature at Bottom of Sheet
T_a = Temperature at Top of Sheet
t = Freezing/growth time
S_i = Salinity of Ice

1. INTRODUCTION

As we all are aware, a special breed of ships is used for icy water regions. Let's have a brief insight into the technicalities of this type of vessels and also delve into their subtle differences and similarities.

1.1 TYPES OF ICE SHIPS

Figure 1: MS Viking Finish Swedish Ice Class Vessel (Courtesy: Wikipedia)

The rules and guidelines incorporated for the construction and construed for classification are known as the "Ice Class" Rules. Let us briefly contemplate into the detailed nuances amongst the broadly classified types:

Ice-Strengthened Ships: These have additional 'ice-strengthening' and ice capable performance which qualify it to move in icy waters. Again by ice-capable performance, we mean adequate engine power rating, specialised propulsion, machinery, high-end material provisions in areas of higher vulnerability etc. They require an escort of icebreakers while moving in ice as they are not self-sufficient to break or plough through ice especially in more drastic conditions. They do not have significantly altered hull design and are often characterised by a common bulb form bow (bulbous bow).

Ice-going vessels: They are a higher breed of ice-strengthened ships not having the requirement on an icebreaker escort. Their hull form, build, propulsion, navigational features and most importantly, strength are incorporated to exclusively function in icy waters.

Icebreakers: They are specialised vessels with the sole purpose of breaking through all kinds of ice conditions and escort other vessels as discussed above. Goes without mentioning that they have all high-end features of 'Ice-Class' often dubbed as the Polar Class. An icebreaker basically makes a channel after breaking through the ice cover. This makes way for other vessels to move by. It has been observed that the width of the ploughed channel is roughly two ice thicknesses more than the ship's beam.

Figure 2: Russian Nuclear Icebreaker Yaman (Courtesy: Wikipedia)

All three of these ships have varied similarities and differences. However, the principal purpose of all of them remains the same in varying degrees: to negotiate and sustain heavy moduli of ice loads without affecting the structural strength and functionality of the vessel during operation.

But before delving deeper into the technical and construction aspects of these ships, let us briefly have an insight into the ice environment and the physics of behaviour of ice with such vessels.

2. ENVIRONMENT AND CLIMATE CHANGE – AN IMPORTANT FACTOR FOR ICE CLASSED SHIPS

Though during wintry months, many of the sea routes in the temperate and frigid zones get laden with ice cover, a lion's share of ice cover is concentrated in the polar regions, i.e. the Arctic and the Antarctic.
As per a recent study data, the ice cover extents of the Arctic and Antarctic are $5.1*10^6$ km^2 and $18.2 * 10^6$ km^2 respectively. Due to Global Warming and climate change, this ice cover extents are gradually dwindling.

Apart from the ubiquitous problems of flooding, loss of vegetation or drastic climate change, this poses other problems as well. The broken ice floats farther as icebergs, sheets or floes and has a crucial effect on the navigational traffic of related seas. Worth mentioning, accidents as well due to collision damage with icebergs or icecaps. Remember Titanic?

Thus, the implications of icing have an indispensable presence in the guidelines and regulations of designated vessels for such areas. But first, let us have a brief and simple discussion related to ship-ice interaction.

Ice, as we know can be of various types such as floes, sheets, cakes, bergs or brash ice. Moreover, they differ in their physics, behaviour, strength, and stability. This consequently can have varied impact on their interaction with vessels and other structures.

3. TYPES OF ICE

To sum up, broadly there can be the following types of ice commonly found in the sea [1]:

- Frazil Ice
- Brash Ice
- Floe Ice
- Ice Cake
- Fast Ice
- Pack Ice

Figure 3: Broken shards of new ice floating in water (Courtesy: Wikipedia)

Other types do exist, but these are mainly considered in the interest of ice-plying vessels and are related to these principal types only in some way or the other.

Pack ice is large chunks of floating ice triggered from larger ice forms or masses covering a large section of water.

Frazil ice is like ice flakes or needles suspended in water. They have an insignificant effect on vessels but may be considered for resistance and abrasive effects onto the hull.

Brash ice is in the form of loosened fragments, basically, like a wreckage of bigger forms of floating ice lesser than 3 metres thick and 2-metre diameter.

Floes are basically flat pieces of ice sheets spread over a vast expanse with small variations in thickness. They too are larger break offs from pack ice.

Cake ice, as depicted from the name, are circular pieces of ice with a certain thickness welling up from the mean sea level.

Fast ice is harder forms of ice, which generally remain attached to the shore or ice wall. They are often dictated by surrounding sea level and may replenish themselves over years.

The age of an ice is a foremost factor. Here too, they may be classified into following main kinds of heads:

- *New ice and young ice*
- *First-year ice*
- *Old ice*

New ice is formed as a result of the recent freezing of sea water. They have less thickness (up to 10 cm) and do not pose very hazardous ice loads on ships. But they are never to be neglected in ice class designs. They include frazil ice, brash ice or cakes.

Young ice is the slightly harder and thicker (10 cm-30 cm). If they survive up to one year, they may accrue into ***First Year Ice*** which typically have thicknesses varying from over 30cm up to 200cm.

If a pack of ice survives multiple melting seasons, they may accrue into a harder form. This is the **multi-year ice or Old Ice**. Another factor is the size of ice. The expanse of an ice cover varies from sporadic free-floating pieces to large packs and ridges.

Ridges are said to be the thickest sea ice features and account for half of the total ice volume. They are mount-like features formed as a result of continued stress, Convergence and collision between two successive ice covers. Found gradually on the sea ice boundaries, they consist of layers of ice rubble and floes piled up above and below the waterline. The part below the waterline is generally larger in size as compared to the part above. Nonetheless, all forms of ice pose threat in varying degrees to vessels plying on icy waters and needs to be taken care of. So, in the design regulations of these vessels, several factors are taken into consideration, such as:

- The expanse of ice cover
- Level ice thickness and floe size
- Sea and air temperatures
- Density and Concentration
- Age of ice cover, as described above

4. SHIP DESIGN FOR ICE

Prior to speculating these aspects, let us first look into the fundamentals of ice loads.

4.1 WHAT ARE ICE LOADS ON SHIPS?

Like traditional hydrostatic and hydrodynamic loads, the ice too poses high modulus of forces onto the hull as obvious. Moreover, ice being in a solid state and having varied sizes and features can cause damage to the hull if unattended. So, vessels plying in icy regions have different build specifications taken into account.

Now, think of this? When a ship encounters ice, any or a combination of the following may occur:

- The vessel may ram or stave through the ice
- The vessel may drift the ice sideward if given suffice space and low interaction forces
- The vessel may climb the ice and press it downwards by the virtue of its weight
- The vessel may shear through the ice, breaking it into smaller fragments henceforth.

As the age, formation, and size of ice in a particular geographic region are multi-variant, the vessel must be structurally and functionally capable of withstanding all plausible types of ice conditions. Apart from the main hull, the propeller and rudder along with appendages (if any) must be capable of enduring ice interactions without compromising the required design performance of the vessel. Continuing our discussion with icing loads, we can look into 3 crucial aspects of ice loads. They are:

- Physically encountered ice conditions
- Statistics of ice loads

- Mechanics of ice loads

While the encountered ice conditions deal with the physical properties, distributions, occurrences, and variations of ice, the statistics include dealing with statistical data of climates and icing. Mechanics is the end result of these as they directly dictate the extent of design philosophies incorporated into a vessel enduring such conditions.Now, when we contemplate about vessels plying in ice, the most crucial thing that logically comes to the mind is the need for strengthening. This is obviously true. Classification regulations for strength are incorporated in all sorts of such vessels in varying degrees. From the design point of view, some other alterations are also made depending on the type of vessel and its utility.

[2]As we can understand, the design performance of an icebreaking ship is decided by the ships' forward speed in these given ice conditions. Moreover, the structural integrity of the vessel should not be compromised. So the foremost aspect that comes to our mind is the incidence of 'ice loads'. From the hydrodynamic analogy, the ice loads too can be segregated to:

- Global Loads
- Local Loads

Local loads can be described as loads coming from single-site ice contacts at multiple locations or a fraction of a larger loading on a particular hull element. For instance, if we have a cusp of ice contacting a part of a hull, then it's a local load! Or if we divide a larger contact force acting on a particular plate panel on a certain frame during a certain span of time, then that's again a local load. This can be depicted in the figure below.

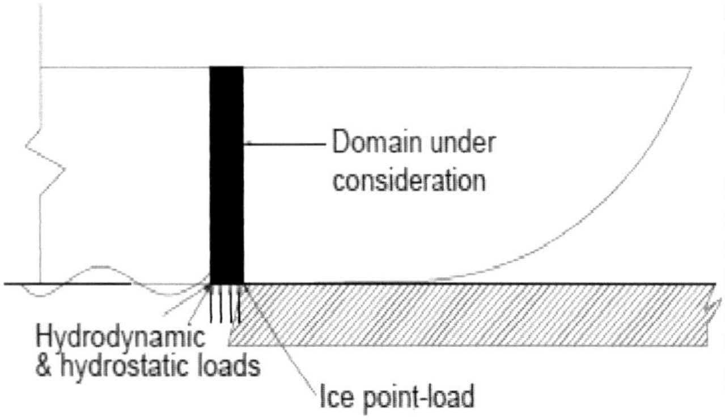

Figure 4: Depiction of localised ice loads acting on a ship

On the other hand, the net load from an ice cover acting on the larger elements like the entire hull girder or say, the side shell plating longitudinally is the Global Loading. To put it simply, the sum total of various forces simultaneously acting on the entire ship is the **global load**. This is the most crucial aspect for a designer in terms of selecting the hull form, structural characteristics, build specifications and the build material. Like hydrodynamics, global ice loads consider the hull girder as a single simply supported beam enacted upon by distributed loading from ice and water. Basically, the load response on the hull are time-variant reaction forces which occur due to various load incidence processes. These loads are:

- Forces due to crushing/breaking the ice
- Forces due to breaking by buckling/bending or flexural failure
- Forces by submergence of ice
- Frictional forces due to sliding, grazing or breaking of ice fragments along the hull (Sliding Forces)
- Open water hydrodynamic forces

These above are nothing but different failure modes of ice. Ice exhibits varying strength against these different failure modes. These somewhat depend on the nature of ice and impact area. Brash or fragment ice has lesser strength. Also, owing to their physical size, the area of impact is small. Thus they are easily yielded by crushing. Consequently, the impact pressures upon the vessel are less. Breaking or crushing of ice begins when the stem of the ship comes in direct contact with the ice cover.

The ice gets grinded into smaller fragments which get dispensed in the surrounding water. This continues for some time. Then at a certain distance away from the point of collision, the impact forces are transmitted. A similar situation can be visualized like this. Suppose you pelt a piece of stone onto a glass anywhere. At the point of impact and the vicinal areas, the effect is more to form a cavity.

However, slowly tracing the fault, cracks can be observed up to longer extents varying from intense ones to fine hairline ones directly depending on the distance from the point of impact. Now if this sheet is subjected to some continuous force, there is a high possibility of a shear failure at some distance away from the impact and perhaps a consequent breakage.

So happens in case of ice sheets. The increasing distance from the point of impact creates linearly increasing bending moment that leads to gross shear failure after a certain point. Obviously, this is the case mostly for more expansive ice covers or bergs whereas smaller fragments are mainly yielded by crushing.

Often the vessel may climb over an ice sheet during the interaction process. As a result, the bending failure happens partially owing to the vessel's weight. This leads to submergence loads.

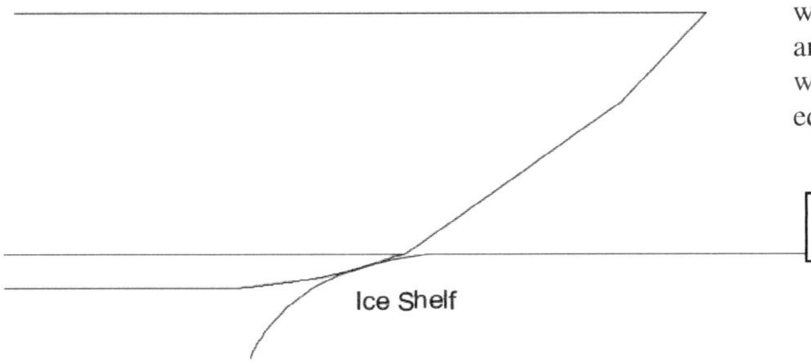

Figure 5: Climbing over an Ice Shelf

Friction is a ubiquitous aspect. Ice is definitely not an exception. The grazing, rubbing, collisions and breakage of ice pieces in various degrees create significant modulus of frictional forces. These enact locally but aggregate to give rise to resistance forces affecting the overall surge. Not to forget the 'localized' impacts on the hull plating due to their continuous interactions. Also, last but not least, open water hydrodynamic loads that are always persistent.

Friction from ice is an integral part of ice resistance about which we shall delve at a later course. But at this stage, it suffices to state that the surface properties of ice affect the 'abrasive' properties, which in turn impact the friction forces enacted upon by the incident ice and consequently, **Ice Resistance.**

Now obvious physical properties of ice come into play like the strength of the ice, thickness and density. These basic physical properties dictate the nature of ice, which in turn affects ice loads and consequently, resistance.

4.2 PHYSICAL PROPERTIES OF ICE AND THEIR EFFECTS

Before discussing resistance, let us look at how these physical properties affect ice behaviour and their induced loads.

4.2 (a) Ice Thickness

Ice thickness, as seemingly obvious in determining the strength and thus the ice loading. This thickness, is in turn, determined by a number of factors such as freezing time, age, ambient temperature, snow type, wind speed etc. More the freezing time, lesser the ambient temperature, lesser the wind speed (due to its wearing effect), the thicker the ice grows. Stefan's equation gives a relation for thickness:

$$H_i = (2*k_i / (\rho_i *L_f)* (T_b - T_a)*t)^{(1/2)}$$

4.2 (b) Density

With the word density, the foremost thing that comes to our mind is weight coupled with compactness. This influences the impact force and also the *buoyancy*. The buoyancy of icy waters has been a complex phenomenon. Apparently, as the buoyant force is provided by water itself, the role of ice is often neglected. However, as ice is much denser than water, a smaller fraction of icy water can provide the required buoyancy force equivalent to a larger volume of full water. Also, in instances when the vessel climbs up a sheet of ice, the lift forces accounted for is quite significant. This is interrelated with two other very significant parameters, salinity and temperature.

4.2 (c) Salinity

Salinity is the ratio of the salt content in ice versus the mass of ice. This increases the density as expected. Another factor comes into play, namely the temperature. Observations reveal that with increasing salinity, density is more sensitive to temperature changes. The physics is simple. Higher salinity implies more brine content, more sensitive to temperature changes. Salinity is inversely proportional to the ice thickness as $S_i \propto 1/h_i$. Thus multi-year ice have lower salinity than new ice.

4.2 (d) Porosity

All ice have an inherent porosity characteristic. This is tantamount to the amount of gas trapped inside and brine content, affecting its rigidity and strength. This is directly proportional to the salinity and inversely to temperature for obvious reasons. Furthermore, the size and distribution of the pores also affect the resultant strength of the ice cover under consideration.

4.2 (e) Flexural Strength

Probably it is one of the most crucial physical aspects of ice. Flexural strength is the measure of the ice's ability to bend before failing. As the name suggests, flexural strength is the inherent material property, conceived as the maximum permissible stress sustainable by the material before it ruptures due to bending loads. This flexural strength is in turn, directly proportional to brine fraction. So, ice in more salty waters tends to be stronger! Normally, it is found to be between 0.1MPa to 1.5 MPa roughly.

4.2 (f) Concentration

In an expanse of icy water, the concentration of ice cover plays a crucial role. The lesser the concentration, the easier is the manoeuvre and performance. So, concentration is expressed as a ratio of ice cover per unit surface area of water normally expressed in a fraction out of 10. A classification 'Egg Code' is also deployed in dealing with the same.

4.3 BREAKING PROCESS OF ICE

Now the main breaking process of ice is an interplay of these various physical properties in the incidence of the stem of the vessel ramming through an expanse of ice. The most commonly encountered ice condit' is 'level ice'. Though other several ice forms e like ridges, floes or pack ice, the most conven: way to depict the ice-breaking process is thro level ice. This can be very well illustrated in following figure:

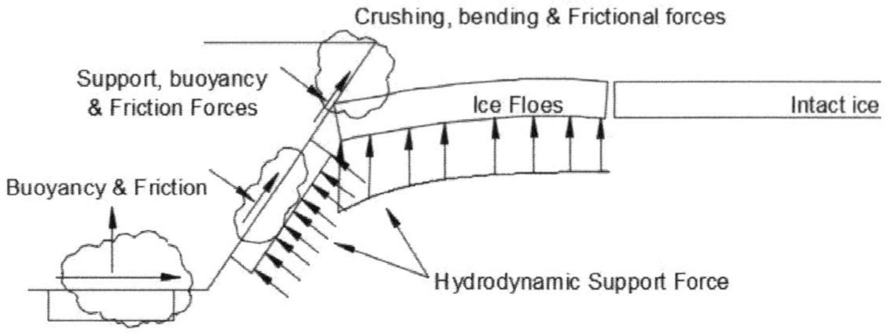

Figure 6: Representation of Ship-Ice Interaction

In the initial stages, when the ice sheet is being impacted upon, numerous local loads are generated in the stem due to the 'crushing of ice'. Small fragments of ice in the form of cusps and floes are generated. This process continues for some time. Gradually, the ice sheet is being delivered upon by a longitudinal force in terms of flexural loading, in turn, caused by compression and buckling loads. Tantamount to this is the submergence load in case of the vessel tending to climb upon the ice. This leads to huge bending loads on to the ice sheet gradually affecting its flexural strength. As the crushing/ breaking action is already in an advanced stage by this time the surface area increases for the action of the contact force. This increases the vertical component of the contact force of ship-ice interaction at the bow from the simple relation of **Force= Pressure* Area**, the pressure being from the enacting ice load. This coupled with the 'climbing force' of the vessel leads to the formation of a bending crack. After a certain point of time, the ice sheet breaks off at the point of maximum bending load. The other forces acting during this interaction process are local reaction forces (at the point of contact during breaking), friction forces from the grazed off ice floes, hydrodynamic forces and of course, the support and buoyancy forces.s

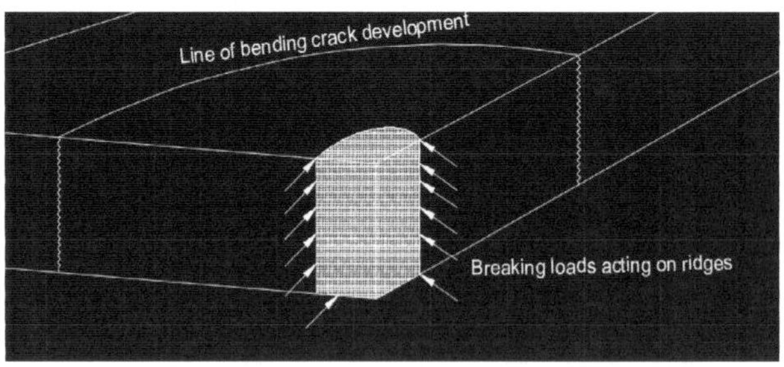

Figure 7: A CAD visualization of ice breakage due to bending

After the breakage, the ice gets disintegrated to floes, cusps, bergs or other smaller fragments. Now, these entities of ice pose a different kind of problem in terms of ice loading, resistance and hydrodynamics.

As per the classification, the design of the hull form plays a pivotal role in defining the same.

Some key factors in the hull form are stem angle, entrance, buttocks, beam, and flare, waterlines. As the main capability of an ice-class ship lies in its ice-negotiating and breaking capability without a compromise of its operational utility, certain crucial parameters are taken into consideration such as:

- Determination of the likely loads (based on ice characteristics, expanse, areas of the voyage, temperatures, ice covers and most importantly, the operational nature of the ship)
- Framing system
- Allowable and limit-strength design values, the overall detailed design (make, dimensions, hull form, propulsion system, power output etc.)
- Hull Material
- Paints and coatings
- Design of propellers and shafting based on the feasible zones of navigation

Ostensibly these are invariably applicable to more or less all forms of ships. However, when plying in ice, the importance of these becomes manifold. For instance, in any other vessel, loads encompass principally hydrostatic and hydrodynamic loads along with secondary ones like wind or weather loads. From a design point of view, a vessel plying in icy regions is subjected to ice in the lower portions of the hull as obvious. So, needless to say, that only those areas of the hull are given special consideration in terms of ice-strengthening.

5 ICE RESISTANCE

Before carrying out our discussion further into the design and classification aspects of ice-worthy vessels, let's us briefly touch upon the crucial aspect of ice resistance. As we know, ships already are encountered with frictional, wave resistance and form drag in open water along with other smaller addendums like an appendage or air resistance. Of course, it goes without saying that in icy waters the ice poses a value of resistance which irreversibly affects the propulsion and powering problems of the vessel.

To a layman, the resistance imposed by ice may appear to be akin to the ones posed by the road to a moving vehicle, which in elementary physics is simply coined as 'friction' force. But here it is more complex. The resistance encountered by vessels in icy waters is an interplay of various phenomenon exhibited by the intact or broken (brash) ice during its interaction with the hull.

As we have discussed in our previous versions, when a vessel encounters a sheet of ice in its course, it rams into it breaking it off to floes, fragments, ridges etc. The aggregate ice resistance is comprised of the resistance due to:

- *The crushing or breaking of ice, which can be visualised as more of a mechanical resistance*
- *Turning/drifting and moving of the ice floes sideways*
- *Resistance due to shear failure of ice*
- *Submergence of the broken ice.*
- *Level Ice resistance*

Let's begin with the last one. Level ice resistance is the simplest form of ice resistance. Suppose, the icebreaker or an ice-worthy vessel is stuck abaft or affront an expanse of level ice or simply an ice sheet afloat in water.

During this process, the stem of the vessel will tend to interact with the ice cover for the purpose of breaking it. The stem of the hull imposes a pushing force tending to barge through the ice which in turn triggers high moduli of reaction force from Newton's 3rd law.

This will provide a considerable degree of resistance to the ship tending to move ahead or astern (if the ice sheet is behind) to the effect citing the interplay of the hull contact and the reaction forces. The situation can be best compared to a tug towing a larger vessel. The maximum towing force occurs at zero speed (when the towing line is fully taut and the larger vessel just begins to move), also known as the situation of ***maximum bollard pull***. So, in this situation as well, the pushing force (analogous to towing force) value oscillates from zero (at the zeroth time) to a maximum constant value (after some time, when the ice sheet just begins to crush).

So the net ice resistance, in this case, is measured as the time-average value of the 'pushing force' (like the towing force of tugs). Maximum ice resistance (static) occurs at zero speed and is higher than at low speeds (when the vessel just begins to move through the crushed ice).

As the ice gradually starts breaking, the velocity of the vessel increases as expected along with higher applied thrust and the 'ice resistance due to velocity'/ dynamic resistance increases linearly with velocity. An obvious physical property dictating the level of ice resistance is the thickness of the ice.

Resistance during crushing can be reckoned interchangeably with the Resistance in Level Ice. Just an obvious difference is that resistance due to crushing encompasses all forms of ice including uneven ridges as well. So, the resistance force in level ice can be considered as a subset of the resistance categories due to crushing. Here, resistance due to ice is principally due to the time-variant horizontal component of the reaction forces during the ship-ice interaction. This is well depicted in the following figure.

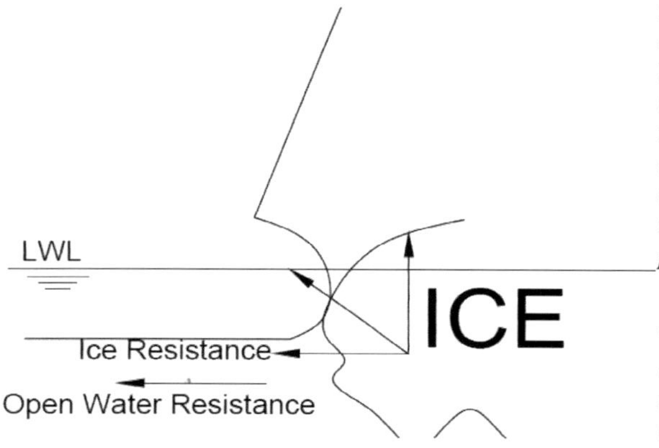

Figure 8: Breakup of Resistance in Icy Waters

The resistance component due to shear failure follows local crushing resistance in most cases. The ice-capable design is aimed at inducing higher moduli of vertical components of contact force.

This breakage of ice at a distance away, as we know can be attained in two ways: One by constant contact force, wherein the ice may fail due to 'buckling' (due to compression at the edges) or 'bending'. Another by partially 'crawling' or climbing the ice cover and breaking it by 'bending'. The disparity in terms of bending and buckling lies in the angle of interaction and friction (between ice and the structure).

The second probability of the first case and the second case is more common. As we had already discussed, the application of compressive or pressure loads, in this case, increases flexural stresses due to the reaction force of the ice which is somewhat 'hydraulically elastic in nature'.

Now an obvious question pops up: How can the ice be brittle exhibit elastic properties? The answer lies in the expanse of ice cover. Long sheets of ice over a region have a certain extent of hydro-elastic behaviour analogous to a strip of metal!

This aspect makes its fail by buckling and predominantly, bending. However, we omit to delve deep into the complex mechanics of ice elasticity. **That's why in larger ice covers, crushing is not that important**.

In both cases, there cracks propagate radially until the point of failure due to flexural stresses. Flexural stress values increase and after sometime approaches the limit stress of bending. This is dependent on the contact area as a larger contact area subjugated to continual loading over time leads to shear forces. This is visualized by the formation of radial cracks at varying distances from the point of application. Spontaneously, the local stresses increase reflected in the form of what we had contemplated the crushing resistance. Due to excessive lateral loads, after a certain tenure of crushing, the ice slab may break/fail at a certain distance due to shear failure, remember? The breakage of the ice sheet appears at a certain concentric radial crack a definite distance away from the point of contact. ***This creates the shear resistance, which is ten times lower than the resistance posed due to crushing!***

Thus, once ice fails due to shear failure, the resistance is decreased manifold. To put it simply, the ice resistance during the ship-ice interaction increases linearly (predominantly crushing till the point of shear fracture and then gradually decreases till reaching a constant value when the resistance is predominantly resistance due to shear failure and the crushing failure diminishes).This is illustrated below:

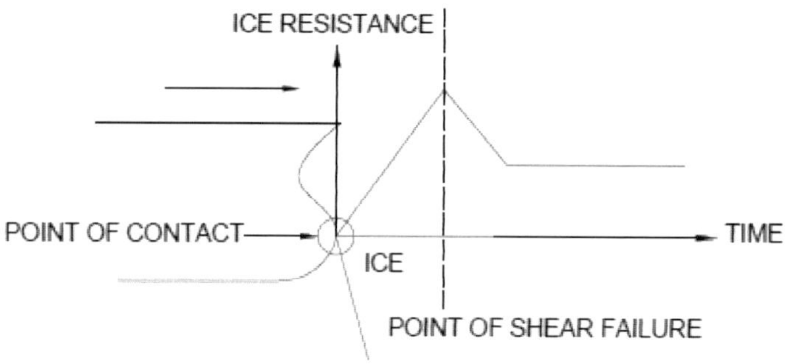

Figure 9

Now after the breakage of ice, the remnant components of ice resistance are posed by the submergence of ice fragments, turning/drifting of ice sideways and sometimes just the movement of ice floes in opposing directions.

Resistances due to turning or lateral movement of ice and submergence are mainly triggered by the simple grazing of the ice against the hull plating creating mechanical friction.

The effect is more pronounced in these two phenomena as the under hull region and the side plating, being the most prone zones of ice contact during submergence and turning, respectively. As understood, the underhull part is affected by the reverse grazing of the ice bits moving aft after submergence, whereas for the side shell it is the constant grazing, impact and collision of the ice floes. The fate of the broken off ice largely depends on the geometry of the hull, which we shall discuss in the next section. Then the ice fragments disintegrate after the interaction with the stem pressure of the hull they tend to come in contact with the side. This is also aided by the buoyant forces and the hydrodynamic forces where these ice floes tend to cling to the hull plating.

Here the surface of contact is mostly flat or even as opposed to the stem where the sharp minimum area tends to wield high pressure required for crushing of the ice. So, the ice follows the contours of the hull amounting to the resistance due to grazing or rubbing or sliding. This ice may now recede further sideways or move underneath the hull depending on the hull form and ice behaviour.

The broken ice can also be set into a rotational motion causing resistance beyond the scope of this article. However, it can be worth saying that the hull form of the vessel and the resultant surface area is of paramount importance as the frictional resistance components of ice are all dependant on this.

Another way of breaking up of ice resistance can be as $\mathbf{R_I = R_B + R_V}$, where R_B and R_V are breaking and speed-dependent parts of ice resistance respectively. The speed-dependant part is again divided into submergence and grazing, as both of them involve a relative motion between the vessel and the ice surface. [3][4]

6. DESIGN PHILOSOPHY OF ICE CAPABLE SHIPS

Discussing the history of all those developments is time-consuming and may be referred to in trusted resources if required. But the crux of designing a suitable hull form for ice ships lies in the aim for reducing the overall ice resistance while effectively breaking and clearing ice.

As we know, ice resistance due to crushing is several times higher than due to bending by shear loads. So, the main aim is to reduce this 'crushing' resistance as much as we can. Also, the open-water resistance (mostly frictional) is also not compromised. So over the years, from experience and theoretical analysis, the hull form of ice vessels have evolved variedly.

Factors taken into account for the design of ice-capable ships are as follows:

- Ship's beam
- Ship angles
- Hull condition
- Ice condition
- Velocity
- Service of the vessel

Ice has a varied response against various kinds of failure modes like crushing, bending and buckling during interaction with the ship. So, the hull form is made firstly to negotiate the ice loads effectively and secondly, the hull is strengthened to counter the ice loads response during various conditions of loading without structural failure.

Ice loads affect both the hull locally, affecting local structures and inducing high-pressure localized 'point' or 'concentrated' loads whereas global loads affect the hull structure as a whole. Modern-day simulations have enhanced the scope of computing plausible loads enacting on the hull girder as well as the local zones of the ship throughout. Sometimes, small-scale model experiments also help. But the proper strengthening of the hull materialistically and structurally is important, but not sufficient. The accurate creation of the hull form is also equally important.One simple thing is that response to resistance must not be confused with loads, though both are intricately related. If the problem of resistance can be solved, that does not mean the problem of oncoming loads is solved and vice versa. Selection of suitable material and strengthening means improves response to loads up to a certain extent.

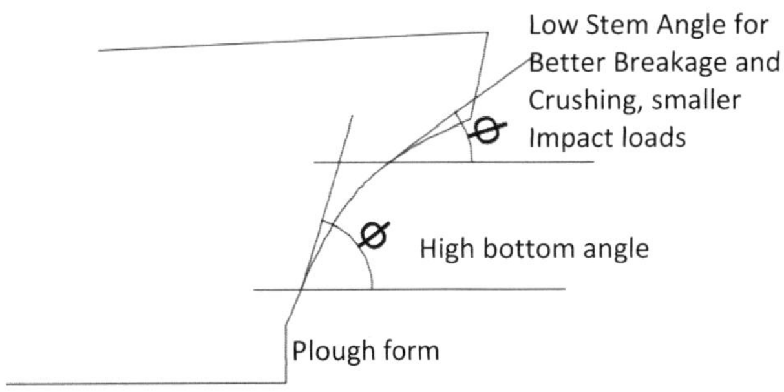

Figure 10

At the end of the day, the complete hull form decides the effective response of the hull to loads as well as resistance. Achieving an ideal design with maximum effectiveness is difficult to achieve. However, data on past designs have made designers informed about the pros and cons of a particular design, making room for improvisation or recreation.

First and foremost, the bow of the ship is of great concern as it cuts or rams through the ice, disintegrates it and clears it to make a channel for ice transition. So, it goes without saying that in most of the modern designs the prow of the ship if of a sharp-edge 'V-form' to enhance cutting through ice. [5]

As breaking the ice by shear to create low-value shear resistance is of main importance, the Stem Angle is made as low as possible as it has been seen that a high stem angle increases crushing resistance which can dramatically increase the total ice resistance. A sharp rake, i.e. a shallow stem angle is found to enhance the vertical component of the enacting ice load by creating a lift of the vessel causing large bending moments amounting to the desired shear resistance more significant over high modulus crushing resistance. This is common in Icebreakers which are meant for high concentration ice covers and have the sole aim of ice by bending (flexural failure). Though no hard and fast rules exist for the stem angle to cater for maximum ice clearing efficiency, an angle of 15-20 degrees has proved to yield desired results. The angle is somewhat increased in moderately icebreaking ships (ice-going) to non-icebreaking ice-strengthened cargo ships which also need to possess open water efficiencies and a larger volume to cater for the capacity. They are mainly designed with the objective of crushing through floes and fragments instead of compact sheets like dedicated icebreakers.

The width of the hull in the forward area is also a crucial factor. Ideally, a low L/B ratio with sharper entrance angles is best for minimizing resistance, especially in icebreakers. However, it is not kept too low as it may pose a problem regarding the desired width of channels for trailing vessels. It goes without saying the maximum beam of the vessel should occur near the forward unlike the midship areas in other general-purpose ships. This is to ensure that the ice slabs affront get broken more easily and end up fast aft near the propeller wash, an important phenomenon for reducing ice resistance.

However, care should be taken to avoid buff forms of the hull in the bow region as this can hinder effective breakage of large ice and at the same time also increase resistance. It has been formulated that the ice resistance of ships chiefly depends on ice thickness, vessel speed and beam. Other than a sharp rake, a plough form at the bow is proved to increase breaking ability with floes drifting sideways towards the wash and thicker ice stomping downwards.

A good design would sharp V-shaped brow with a broad beam close to the bow. In this manner, the Longitudinal Centre of Gravity (LCG) should be well forward of the midship as it will create a narrow tapering typical aft body making the chances of ice accumulation pretty less, also helping clear ice in the downwash.

Also, the lines forward of midship should be less fair to enhance clearance of ice and avoiding clinginess to the hull. So, in this manner, the salvaged ice floes end up in the propeller wash or get laterally "pushed" aside from making a channel. Two classic examples are the spoon-shaped bow and the square-shaped bow, also known as the Thyssen-Was bow. [9]These two have proved to be highly successful in mitigating ice with lesser resistance. They have superseded the previously existing hull form of wedge-shaped bow, which had not proved to be very efficient.

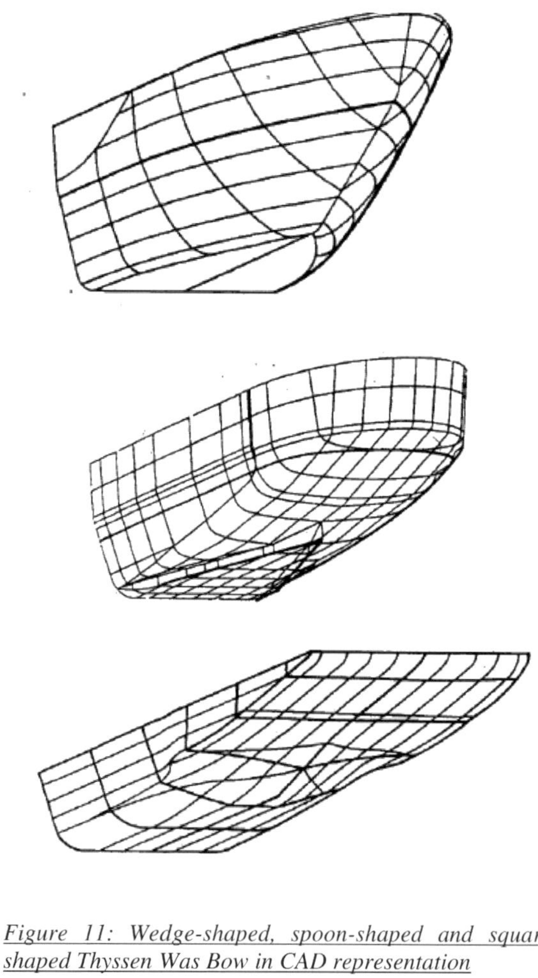

Figure 11: Wedge-shaped, spoon-shaped and square shaped Thyssen Was Bow in CAD representation

The bulbous bow is generally avoided as it has been found to increase ice resistance. Furthermore, since in icy waters, there is a lesser occurrence of waves and the forward speed of the vessel is in nominal limits, wave resistance is not a problem, thus making a bulbous bow absolutely redundant, if not burdening. However, in lower ice-class cargo ships, bulbous bow is often provided as maximum part of its voyage is in open water also providing hydrodynamic efficiencies in channels with small suspended ice fragments.

Square and spoon-shaped bows can increase the shear force for breakage and cause lesser resistance. A circular bow has also been shown to be effective but can cause a problem of slamming problems.

In modern bow designs of ice-capable ships, the wide beam in the forward region rapidly tapering to the bow has been a successful trend, but this can cause a small problem of hull elongation to make up for its deadweight capacity. Lengthy hulls, on the other hand, can pose problems in turning and lateral movement in ice.

The stern of efficiently proven vessels can be envisaged to be sharp, steep and narrow for good seakeeping and icebreaking properties. A stern is often kept very high with more than ample propeller clearance to cater for lesser accumulation or 'clinging' of ice in it's after wash and enhance the creation of a better channel in case of dedicated icebreakers. Stern shoulders are also of importance as avoidance of blunt shoulders can achieve good reversing performance and manoeuvring abilities. Ice-capable ships like breakers and ice-strengthened ones are characterised by the rapid collapse of frames to better deal with broken-off ice by either pushing it aside from the channel or plunge it down towards the propeller wash.

Another interesting feature of ice-capable ships is the presence of ***reamers***, both near bow and stern. Reamers are merge points of a bow and middle body in the forward and stern and middle body at the aft. They are like sharp knuckles which act as obtrusion to ice wedges, pushing them aside the channel underneath the ice wall boundaries.

Stern reamers also similarly break ice aft of midbody and push the ice underneath the channels when operating astern. Regarding waterlines, the half-angle of the entrance is desired to be kept in the same order as the rake/stem angle as the fineness of the waterline is said to produce large downward forces required for bending, thus shearing. Block Coefficients are variable, mainly dependent on the bow and stern forms and the capacity, but is tried to be kept low.

Midship section coefficients may vary but can be kept high owing to sea keeping, stability purposes and often to compensate the lesser available volume in the forward and aft regions owing to its typical hull form. The flare of these ships has always been a debatable topic.

European designers had postulated over the years that a small flare angle would be congenial in lifting the vessels desired for breaking the ice by bending whereas Canadians had argued that a large angle would be equally opportune. No concrete judgement is available. However, proven assertions from U.S Coast Guard and several latest Russian designs have vouched for a minimum 10-20 degrees flare.

Another common technology in modern icebreakers is the presence of bow and stern propellers which help in the clearing, disintegrating and washing off of ice, also facilitating proper flow of broken ice in its downwash.

7. A BRIEF INTRODUCTION TO ICE CLASS

Needless to say that all ice-plying vessels are governed by a different set of classification rules dictating their make, building materials, design, arrangement, structure, functionality, equipment, stability, machinery, safety, etc.

All vessels like icebreakers, ice-strengthened ships and other ice-going vessels are all encompassed under Ice Classification Rules.

Several (but not all) classification societies have their own set of rules under the name Polar Class (PC) which deals with the capabilities of all these ships qualifies to ply in icy waters. Also, a set of exclusive generalized regulations have been adopted by the IMO namely, the '*International Code of Ships Operating in Polar Waters*'. The foremost classification widely followed by more or less all classification societies is the segregation of vessels based on serviceability. They are descended from PC 1-7 based on the service of the vessel, the ice conditions and most importantly, the timing of operations [7] -

PC 1: Year-round operations in polar waters

PC 2: Year-round operations in moderate multi-ice conditions

PC 3: Year-round operations in second-year ice

PC 4: Year-round operations in first-year ice, which may include old ice inclusions

PC 5: Year-round operations in medium first-year ice which may include old ice inclusions.

PC 6: Summer/autumn operations in medium first year ice, which may include old ice inclusions.

PC 7: Summer/autumn operations in thin first-year ice, which may include old ice inclusions

Also, classification societies like DNV-GL have their own set of classification as also do another set of rules called Finnish-Swedish Ice Class Rules. For example, as per DNV rules:

- Ice Class 1A*: Highest class notation alluding to ships operating in all ice conditions, from extreme to moderate, without icebreaker assistance. Obviously, this includes ice-going and icebreakers and definitely not simple ice-strengthened/ice-class ships.

- Ice Class 1A: Capable of trotting in difficult ice conditions, with icebreaker escort.

- Ice Class 1B: Moderate ice conditions with an icebreaker.

- Ice Class 1C: Light ice conditions, with icebreakers. Mostly, cargo ships capable of lying in very cold regions.

Also, IMO's '*International Code of Ships Operating in Polar Waters* also have their own categorization, viz., Category A, B, C. We do not delve into the details of each and every aspects of classification. But a crucial aspect of classification is the material selection and strengthening.

Needless to say, ice-class vessels are given extra care in terms of strengthening and material selection to sustain in their operation. An ice-class defines a region in ship's hull where maximum plausible ice loads can occur and names it as 'Ice Belt' [6]. This demarcation helps in identifying the zone of maximum plausible loads from ice

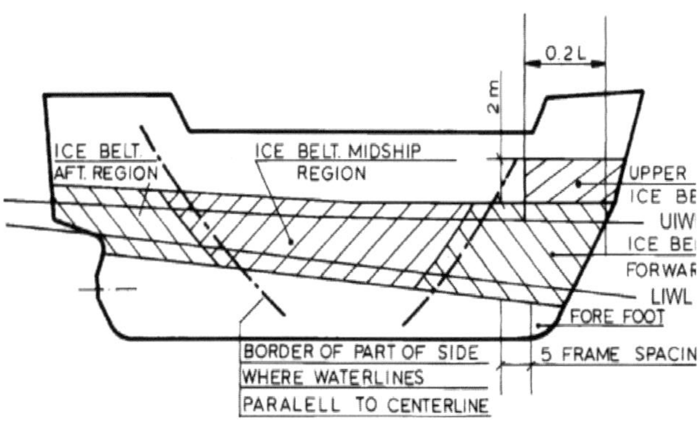

Figure 12: Demarcation of Ice Belt Demarcation of a typical Ice-Class Ship under Class Norms (Source: DNV-GL)

The ice belt is flanked by the Upper Ice Water Line (UWIL) and Lower Ice Water Line (LWIL). The ice belt region is given extra care by the classification guidelines in terms of more strength, scantling and stiffening arrangements. UWIL and LWIL are the upper and the lower limit of the waterlines in icy waters, respectively which are used to estimate the feasible zones of ice loading in ships. [7]

[7]So, in all ice-capable ships, the materials used in the ice belt is high breed high tensile steel of various grades like AH, DH, EH. Other areas can be either high tensile steel or mild steel (B, D, E, F etc.) of varying scantlings as per requirement. The details of all such rules and regulations are available in classification documents. The foremost aspect taken into consideration while selecting the material for hull is the worst-case scenario of low temperatures which can hamper the ductility and drastically increase the brittle nature.

Obviously, the Ice Belt is different in icebreakers and ice-going ships as compared to simple ice class. It goes without saying that in prone areas, the stiffening and framing arrangements are especially decided based on the vulnerability of oncoming ice loads, where the stiffening members like ice stringers etc. may have high scantlings and closely spaced depending on norms and guidelines (often flat bulb profiles are used) . The shell plating is supported by main frames, which in turn are supported by stringers as shown. These are further supported by close web frames and bulkheads. The following figure represents an ice-belt in a vessel and the arrangement of stiffening members.

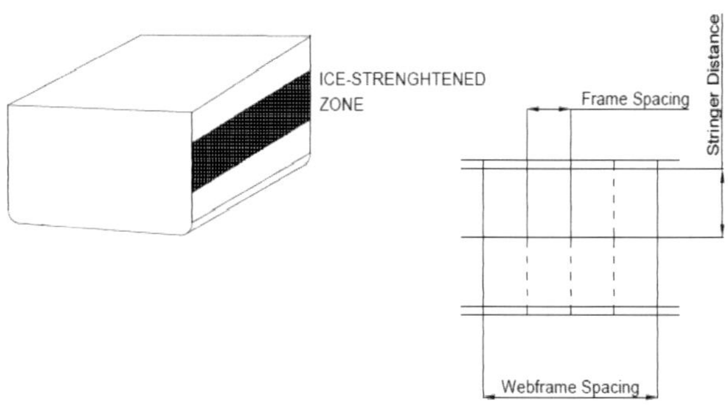

Figure 13

Corrosion and abrasion protection are also given as per classification guidelines as the hull shell plating is highly vulnerable to corrosion, freezing, embrittlement and abrasion in frigid conditions. The most common choice for propeller are high-grade stainless steel and Nickel-Aluminium Bronze.

8. CONCLUSION

With the ever-increasing traffic in cold ocean routes, the call for incorporation of ice-class design to more and more vessels has gained paramount importance. Moreover, the discovery of natural reserves and the increasing popularity of oceanographic research in extreme ice-capped regions has seen the deployment of ice-breakers and ice-going vessels in the most hostile environments. Like other ships, these breeds of vessels have undergone evolution in design over the years and still have a long way to go. With latest guidelines in aspects like EEDI, pollution and an insatiable optimization towards operating costs, the improvements, research-based developments and re-designing of these crucial vessels will continue to be a quest for all naval architects.

9. ACKNOWLEDGEMENTS

- Mr. Raunek Kantharia, Co-founder and Chief Editor, Marine Insight
- Mr. Bijit Sarkar, Naval Architect
- Mr. Tatan Paul, Lloyd's Register Asia

10. REFERENCES

1. J. DOLNY, 'Methodology for Defining Technical Safe Speeds for Light Ice-Strengthened Government Vessels Operating in Ice ', *ABS Ship Structure Committee*, Nov. 2016.

2. 'Structural Challenges faced by Arctic Ships', *Ship Structural Committee*, 2011

3. T. SKAR, 'Ice Induced Resistance of Ship Hulls', *NTNU*, 2011

4. J. LEWIS, F. DeBORD, V. BULAT, 'Resistance & Propulsion of Ice-Worthy Ships', *SNAME*, 1982

5. K. RISKA, Ship-Ice Interaction in Ship Design: Design and Practise, *University of Science and Technology, Norway*

6. F.B WALLIN, C.J. AKERSTROM, 'Ice Class Requirements for Side Shell Structures', *Chalmers's University*, 2012.

7. MAN B&W, 'Ice-Class Ships Main Engines'

8. GL, Rules and Guidelines for Classification of Polar Ships, 2008

9. J. RYBKA, *SNAME*, 2001

11. AUTHORS BIOGRAPHY

Subhodeep Ghosh holds the current position of Design Engineer at Reliance Naval and Engineering Limited, India. He is responsible for basic design work of defence vessels and crafts. He is a graduate of Naval Architecture and Ocean Engineering from National Ship Design and Research Centre, Indian Maritime University and is a freelance writer for Marine Insight.

DISCLAIMER

The given paper is the authentic and sole effort of the author expounding the concepts in simple terms discernible to all kinds of readers. The concepts of the paper had been previously published by the same author as a 4-article series in the e-newletter **Marine Insight**, which had completely agreed towards the content to be reproduced for the Conference held by The **Royal Institute of Naval Architects**. The author is a technical writer for Marine Insight as well. The author safeguards the right against plagiarism or cloning of the content by any other source or individual.

MARINER 4.0: INTEGRATING SEAFARERS INTO A MARITIME 4.0 ENVIRONMENT

N. C. Taylor, **A. Bekker** and **K. Kruger**, Stellenbosch University, South Africa

SUMMARY

Digitalisation of systems and processes in the maritime industry are spurring *Maritime 4.0*, a digital transformation promising enormous gains, including improving design processes and reducing operational costs. Gleaning inter-disciplinary proficiency from Operator 4.0 in manufacturing, the present work seeks to introduce the *Mariner 4.0* – a seafaring passenger or crew member with technologically augmented skills. This research aims to realise Mariner 4.0 through a human digital twin solution – a virtual representation of the state and behaviour of a unique individual over time – for seafarers on South Africa's polar research and supply vessel, the SA Agulhas II. This paper presents human-related challenges identified within seafaring vessels, the expected value of human digital twin solutions for the maritime industry, the development of a human digital twin solution to integrate seafarers into a Maritime 4.0 environment (i.e. the development of Mariner 4.0), and an illustrative use case focussing on human comfort monitoring and management.

1. INTRODUCTION

Operator 4.0 is defined as a human worker that uses technology to augment their skills and abilities to perform work activities more efficiently, creating a smart and skilled worker in the manufacturing industry [1]. With inspiration from Operator 4.0, the present work seeks to introduce *Mariner 4.0* – a seafaring passenger or crew member with technologically augmented skills, forming smart, skilled passengers and crew operating in digitalised maritime environments. Mariner 4.0 aims to embody human factors in digital systems and enable interfacing seafarers with machines and digital systems. Mariner 4.0 strives to represent a human-centric development within Maritime 4.0 by recognising the value of passengers and crew on seafaring vessels and endeavouring to facilitate the inclusion of passengers and crew in digital developments.

Maritime 4.0 can be defined as an integrated system of systems concept in which the implementation of digital processes and technologies is exploited to realise digitalisation throughout the entire value chain of the maritime industry [2]. The objective of Maritime 4.0 is to deliver high quality, optimised and reliable systems that leverage state-of-the-art technologies to maximise customer value [2]. The current phase of digitalisation in the maritime industry could be coined Maritime 4.0, as it focusses on revolutionising the maritime industry through adopting the use of digital technologies and concepts – the definition of a *digital transformation* [3].

The development of Maritime 4.0 is influenced by *Industry 4.0*, aimed at developing industrial production systems [4]. The Industry 4.0 concept is realised by creating smart cyber-physical systems (CPSs) and applying internet of things (IoT) to industrial production systems. The Industry 4.0 initiative focusses primarily on connecting systems and machines in order to quickly adapt to meet global and fast-changing demands.

Integrating people within the Industry 4.0 vision is seen as an important focus [5]. Integrating seafarers into CPS development in the maritime industry to create *human cyber-physical systems* (HCPSs) endeavours to foster human competencies and enhance human-machine symbiosis in Maritime 4.0 [1]. A *HCPS* is a system created to:

- Improve human abilities to interact and communicate with machines through human-machine interfaces, in both the cyber and physical realms. It is important that these human-machine interfaces are suited to unique human capabilities.
- Augment human capabilities by using technology, for example by using wearable devices.

It is vital to note that by creating human-machine symbiosis, technology augments human skills and abilities to perform work activities more efficiently [6]. Technology, in this case, is not intended to replace the skills and abilities of humans.

Beside the benefits, the technological developments in Maritime 4.0 influence the working and living environments for seafarers and can be considered socio-economic and technological disruptions providing challenges, such as rapid changes in workforce skill requirements and employment patterns [7]. Future workforce skill sets may be vastly different from today and may demand abilities that humans cannot develop [8]. A conventional view for driving ship autonomy included removing people on board to reduce human error, but an ethnography of maritime autonomy explains that there seems to be a greater desire to keep people on board and use automation to augment seafarer abilities in order to reduce the workload of crew [8]. The ethnography recognised that humans are likely preventing several accidents that automated systems may not be capable of.

At this stage, people are critical to many seafaring vessel operations and, for example, on cruise vessels human experiences are central to business success. It is important to consider the integration of people within *Maritime 4.0 environments*, which the authors define as an environment that has been influenced by

Maritime 4.0. The proposed Mariner 4.0 shall live or work in a Maritime 4.0 environment and, as such an environment is likely realised through creating CPSs, there remains the need to develop human-centric maritime HCPSs. A potential solution to facilitate human-machine interfaces in a Maritime 4.0 environment, and augment the human ability, is digital twin technology.

A *digital twin* is a virtual representation that renders the state and behaviour of a physical asset, based on sensor observations within its operational context, in close to real-time [9]. In the present work, a seafarer (crew member or passenger on board a seafaring vessel) is considered as the physical asset with an associated *human digital twin* that is a virtual representation of the unique individual, which dynamically reflects the state and behaviour of the individual over time [10]. Developing such a human digital twin for seafarers on South Africa's polar research and supply vessel, the SA Agulhas II, is the main objective of the present work.

Owned by the South African Department of Environmental Affairs, the SA Agulhas II was built a polar class 5 vessel in Rauma shipyard by STX Finland. She accommodates up to 50 crew members and 100 passengers on annual research and resupply voyages to the Southern Ocean and Antarctic research bases. Since her commissioning in 2012, Stellenbosch University and Aalto University have lead a full-scale measurement and research campaign, studying human factors, vessel structural dynamics and environmental conditions during research voyages [11], which this work contributes to.

This paper will present human-related challenges identified within seafaring vessels in the maritime industry, the expected value of human digital twin solutions associated with the identified challenges, the development of a human digital twin solution for integrating seafarers into a Maritime 4.0 environment and an illustrative use case focussing on human comfort monitoring and management.

2. HUMAN-RELATED CHALLENGES ON SEAFARING VESSELS

Human-centric challenges experienced within seafaring vessels were identified in literature and are presented in this section. A few of the human-related challenges encountered may be a result of the digital transformation, but otherwise, technological developments may offer potential solutions as presented in Section 4.

2.1 UNCERTAINTY DURING EMERGENCY EVENTS AND DANGEROUS ACTIVITIES

Uncertainty and high levels of stress, often experienced during emergency events, may influence the safety of seafarers on board. This may especially be the case if safety depends on the ability of significantly dense crowds in confined spaces reaching designated safe mustering areas in a timely and secure manner [12]. During emergency evacuation procedures, knowing the location of all seafarers on board could be critical for efficient mustering [12]. Counting passengers during mustering events can be an impossible task using traditional lists and calling names out; especially when up to several thousand passengers are on board, panicked and there is a pressure on time [13]. It is not uncommon that multi-national crew and passengers are aboard seafaring vessels, which may introduce language and cultural barriers that are especially challenging during emergency events [13].

In the case of missing seafarers during emergency events, being able to find them as a matter of urgency can be challenging; particularly in the case of fire or severe flooding [13]. Using technology to locate seafarers could allow for quicker mustering and efficient searching for missing seafarers [14]. Efforts have already been made to supplement emergency services with passenger and crew location during emergency events, like the Scandinavian Reach service called In:Range [15]. A similar indoor movement tracking system called inFORCE by InnerSpace, that can offer the last known location of individuals, has also been launched [16].

Maritime work may entail dangerous activities, such as underwater welding, which relies on unique knowledge and experience to accomplish. An example scenario, in which uncertainty of surroundings provides unprecedented challenges, is an ice pilot charting routes through relatively unchartered ice environments [17]. In this scenario, existing satellite technology may only provide access to satellite imagery of the ice environment 12 hours retarded. That means today's ice condition "forecast" will only be available tomorrow. The result is that an ice pilot only has yesterday's satellite ice condition information and specialised experience from previous ice piloting to chart through today's continuously changing ice environment.

2.2 CONNECTIVITY OF SEAFARERS AND STAKEHOLDERS

Networked connections, such as satellite connectivity, are developing to such an extent that it is possible for offshore vessels to have connectivity for all on board to reach people onshore [18, 19]. This increased access to Internet offshore also introduces social challenges, for example, more seafarers isolate themselves to communicate with family and friends onshore [20]. Connectivity is, however, in some cases still limited in reliability, speed and bandwidth, due to availability of satellites en route [21], or imposed restrictions, such as no video calling, and high costs [18].

Communication between vessels in low-coverage areas and onshore remains a challenge [21]. Radio communication works well within line-of-sight, but ship-to-shore is not always within range for direct radio communication [22].

2.3 DATA-DRIVEN CHALLENGES

In the maritime industry, the digital transformation has been practically enabled through the development of sensor and data-driven technologies. The development of such technologies has also facilitated increased and easier access to sensors and IoT devices. The maritime industry has seen the implementation of several sensor networks and systems that were opportunity-driven, rather than needs-driven [23]. This has resulted in bulk data capturing without the appropriate data handling methods to extract actionable information and insight, such as design and operational decision support. In many situations, sensors have been placed in suboptimal locations and record large quantities of data that have potentially poor accuracy and fidelity, and at arbitrary frequencies [23].

2.4 LACK OF HUMAN-CENTRIC DIGITAL PLATFORM DESIGN

Digital services use digital technologies to enhance situational insight and support decision-making for improved operations [23]. It seems evident from literature that the maritime digital transformation has focussed on development of digital service portfolios for smart asset management [24], which result in greater development of the digital presence of inanimate assets than the development of human digital representation. Though platforms or systems to monitor vessel or structure health exist, to date there is no recorded human-centric digital platform for monitoring the well-being of seafarers. For example, the automatic identification system (AIS) is primarily focussed on communicating vessel position, course and speed as a safety measure [22]. Nowadays, AIS is used for more than only a safety measure, such as for fleet monitoring, but AIS data does not include seafarer state or well-being. Safety and wellness of seafarers should be a driver for creating human-centric digital platforms that connect vessels, structures and stakeholders within the maritime industry [21].

2.5 FUTURE SKILLS REQUIREMENTS FOR THE MARITIME WORKFORCE

The maritime workforce can be expected to perform laborious tasks with ever-changing schedules and long working hours [8, 25]. The technological developments in Maritime 4.0 will require more technical knowledge and expertise from the future seafaring workforce [26]. These developments will also change the employment patterns in the maritime industry, such as decreasing crew numbers due to increased automation. These potential changes introduce challenges, like the global shortage of skilled seafarers and predicting the future skill requirements of seafarers with the technological advances. These challenges will require continued investment in training to respond to new skills requirements resulting from the technological progress.

Cicek et al. [7] analysed future skills requirements for the maritime workforce with the integration of new technologies in the maritime industry. Important capabilities with regards to four competency categories were found, altogether containing thirty three competencies, including:

- Technological competencies – information and data processing and programming proficiencies.
- Social competencies – ability to transfer knowledge.
- Methodological competencies – information technology and technology affinity, cognitive and learning ability.
- Personal competencies – environmental and safety awareness, flexibility, motivation to learn.

The future maritime workforce may even require abilities that humans cannot develop [8]. This further motivates a human-centric technological development to facilitate human-machine symbiosis within Maritime 4.0.

3. THE HUMAN DIGITAL TWIN

The development of a human digital twin concept for integrating seafarers into a Maritime 4.0 environment, which contributes to the development of the Mariner 4.0 vision, is presented in this section. The human digital twin concept is explained in the context of the present work, also including how to generically model the state and behaviour of a person, followed by the interdisciplinary proficiency gleaned from Operator 4.0 in manufacturing.

3.1 DEFINITION OF THE HUMAN DIGITAL TWIN

A *human digital twin* is a virtual representation of a unique individual, which dynamically reflects the state and behaviour of the individual over time [10]. The term digital twin should not be used synonymously with the terms *digital model* or *digital shadow* [27]. A *digital model* is a static digital representation of a physical object that has no automated data flow between the digital and physical object. A *digital shadow* is an automatically updated digital representation of a physical object that mirrors the behaviour of the physical object. In terms of data flow, a digital model could be considered the lowest level of integration between the physical and digital realms, surpassed by the digital shadow. The highest level of integration is a *digital twin* that integrates automated bidirectional data flow between the virtual and physical objects. The digital twin may provide informational feedback from the cyber world to the real world or, for example in applications of remote commissioning of systems, drive the physical twin.

With reference to the different levels of integration by Kritzinger et al. [27], if a human digital representation only captures the unique attributes of a person, the result may be likened to a *human digital model*. A human digital representation that is a human digital model *and* mimics the state and behaviour of a person may then be a *human digital shadow*. In the context of a *human digital twin*, the digital representation should capture suitable attributes of a unique person and dynamically reflect their state and behaviour over time in such a way to deliver value [10], and provide feedback from the cyber world to the person that augments their ability to work, such as helpful instructions. The human digital twin concept discussed further could be accordingly implemented at the level of human digital *model, shadow* or *twin,* depending on the level of integration required for a specific application.

A generic human digital twin implementation for seafarers on board the SA Agulhas II is shown in Figure 1. Relating the definitions of a digital twin by Erikstad [9] Kritzinger et al. [27] to the present work, in Figure 1 it can be seen that:

- a seafarer is considered as the "physical asset",
- that lives and works on the SA Agulhas II, which is the "operating context",
- virtually mimics the desired state and behaviour of the seafarer, and
- bidirectional communication between the physical and cyber worlds exists.

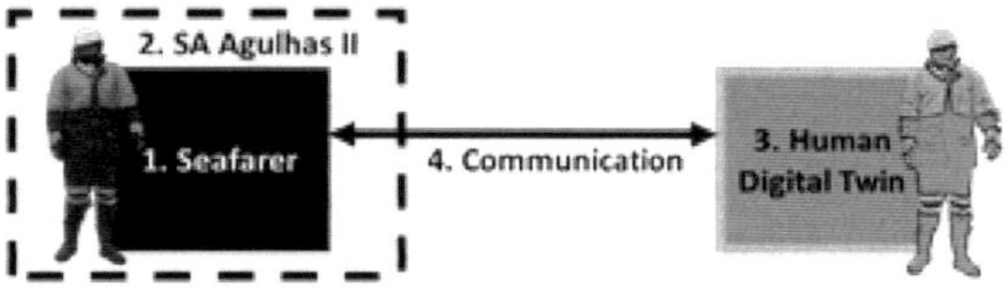

Figure 1: A Generic Human Digital Twin Solution on the SA Agulhas II

The lifecycle of each human digital twin can vary depending on the responsibilities of the seafarer. For a crew member, it may be birthed at the start of training in service, move with them between jobs and eventually retire with them, potentially providing valuable experience for training new seafarers. For passengers, the existence of the human digital twin may only last for one cruise or contribute towards a passenger loyalty service.

3.2 THE HUMAN DIGITAL MODEL

The Operator 4.0 vision considers *human-in-the-loop control* where human behaviour models are incorporated into standard feedback control of systems [1, 28]. One significant challenge is modelling human behaviours due to the complex physiological, psychological and behavioural aspects of human beings [28]. Creating technological systems that are human-centred, such as HCPSs, is possible using human factors concepts and techniques [6]. *Human factors* is the field in which the relationships and interactions between humans and the equipment, processes and products they use is studied. Various human interactions, such as physical, psychological, cognitive and thinking factors, are considered and behavioural sciences are used in the design of machines and human-machine systems [6].

3.3 DEVELOPMENT OF OPERATOR 4.0 IN MANUFACTURING

The shop floor worker has evolved into Operator 4.0 in the course of the fourth industrial revolution. *Operator 4.0* is a smart and skilled worker who works aided by machines when so required [1]. The generation of Operator 4.0 represents human-centric design and engineering, where automation is considered augmentation of the human's physical, sensorial and cognitive capabilities through creating HCPSs.

Romero et al. [29] define Operator 4.0 typologies that capture different means through which technologies can assist workers, such as the *Virtual Operator* that uses virtual reality to aid work. However, the Operator 4.0 vision focusses mainly on augmenting and enhancing human performance through developing human-machine interactions and physical-cyber interfaces. Operator 4.0 defines methods of interacting *with* digital systems, but not yet representing the human in the cyber realm *through* a digital presence. Creating a human digital presence *through* a human digital twin has recently been reported within the manufacturing industry [30].

4. EXPECTED VALUE OF HUMAN DIGITAL TWIN SOLUTIONS

Human digital twins may provide decision support, facilitate the integration of seafarers into Maritime 4.0 environments and contribute to the development of unique human behavioural models, which is expected to offer value to the maritime industry. The potential support human digital twins could offer challenges faced on seafaring vessels is also presented in this section.

4.1 DECISION AIDING

A significant purpose of a human digital twin solution is to support decision-making. In the functional decomposition, presented in Section 5, most functions are dedicated to support making decisions and following through with the decisions made. A human digital twin solution aims to facilitate communication among seafarers on board to augment the capability of seafarers in both making decisions and acting on them. Digital twins in general also provide automated integration of real data into decision-making [2].

4.2 FACILITATE COLLABORATION BETWEEN HUMANS AND DIGITAL SYSTEMS

A Maritime 4.0 environment is realised through the adoption and implementation of connected technologies throughout the lifecycle of assets [2]. This is likely achieved through the design and production of new, smart assets or retrofitting legacy assets with new, smart technology. The working and living environments of seafarers in the maritime industry are experiencing technical transformations as a result of increasing connections between digital and physical systems. Providing seafarers with a digital presence in the largely machine- and system-dominated digital development through human digital twins can facilitate collaboration between digital systems and humans, and machines and humans, which enable them to benefit from digital twin technology in their respective roles on board.

4.3 CAPTURE UNIQUE HUMAN-RELATED INFORMATION

Human digital twin solutions strive to capture unique, relevant information about each seafarer. As an example, human comfort information can be captured through a comfort response (the person submitting a subjective comfort value ranging between 0 and 10, where 0 is severely uncomfortable and 10 is comfortable) during different sea states. This unique information repository can potentially be used in developing personalised models and comfort thresholds for unique individuals. The personalised models developed could be used to shift human comfort monitoring based on past or real-time sensor observations to predicting human comfort for future charters and routes based on sea state forecasts.

Human comfort predictions may facilitate decision-making with regards to planning operations and routes. The most suited crew could be selected based on previous experience and personal responses in similar sea states to predicted sea states, or for training new crew inexperienced in predicted sea states. In more immediate cases, the captain could determine which seafarer may have the most suitable experience in unexpected situations, like finding a doctor among passengers. Furthermore, it should be possible to determine if predicted sea states are fit for safe human-related operations using personalised models to plan routes, perhaps motivating the requirement for autonomous or unmanned vessel operations in dangerous areas.

4.4 POTENTIAL HUMAN DIGITAL TWIN SOLUTIONS TO IDENTIFIED CHALLENGES

Human-centric challenges that could occur on a seafaring vessel were identified in Section 2. The potential resolutions offered by human digital twins to these challenges are presented next.

4.4 (a) Seafarer Location Monitoring during Dangerous Activities

Human digital twins can facilitate location monitoring of seafarers in near real-time, which can be accessible to relevant stakeholders. Immediate access to the last known location of all seafarers can be available during execution of dangerous activities, or to crew and rescue services during emergency events, for example to identify and find possible missing parties. The known location can also be used as a method for controlling unauthorised access.

4.4 (b) Limited Unique and Specialised Knowledge and Experience

The human digital twin of a uniquely knowledgeable and experienced crew member may be able to capture important and critical decisions, as well as tendencies, which could be used in training or augmenting the capabilities of less experienced crew members. Digitising knowledge could potentially be used in development of autonomous vessels.

4.4 (c) Lack of Human-centric Digital Platform Design

Human digital twin solutions are inherently human-centric. A type of digital platform is integral to the realisation of human digital twins in the cyber world, and facilitates links to the real world. A vessel-wide implementation of multiple human digital twin instances can provide a digital platform that facilitates monitoring the well-being of each seafarer on board and augments seafarers' skills and abilities to plan and perform work activities. The platform may be used, for example, to host a single instance of the most-updated vessel information, such as itinerary and schedule, which all seafarers could access through personal interfaces, such as a tablet or smart watch.

4.4 (d) Limited Ship-to-shore Connectivity

The bulk of the data collection and analytics for a digital twin solution could be customised to onboard edge analytics and fog computing, before sharing information about the seafarers directly onshore or to a Cloud platform accessible to stakeholders worldwide. The information could be customised to be only critical or relevant and in compact formats for easier transmission during times ship-to-shore connectivity is available.

4.4 (e) Insufficient Data Management

Need-driven design procedures for implementing human digital twins strive to move from opportunity-driven sensor implementations. The human digital twins could facilitate appropriate data handling methods to extract actionable information and insight, and store organised data.

5. CONCEPTUAL DESIGN OF A HUMAN DIGITAL TWIN

The conceptual development of a human digital twin for seafarers on board the SA Agulhas II, i.e. the development of Mariner 4.0, will be discussed. A systems engineering approach was used, wherein the design process progresses from a broad scope, considering the system as a whole, before being refined into subsystems with specific requirements.

5.1 NEED IDENTIFICATION

Generic human digital twin needs, much like client needs in a standard engineering design process, were determined. Table 1 presents these needs, many of which were inferred from digital twin needs identified in maritime literature [31].

Table 1: Human Digital Twin Needs

Human Digital Twin Need	Need ID
Facilitate insight provision and decision support, at different levels and temporal scales, with regards to seafarer well-being from real operational data. This need further implies automated integration of operational data into decision-making.	N_1
Integration of complex individuals and various sources of data into the digital realm.	N_2
Remote inspection of the well-being, or state, of seafarers and assistance to improving, or managing, their well-being.	N_3
Remote monitoring of seafarer state through controlled access and up-to-date information.	N_4
Platform for reliable bulk data analytics relating to seafarer state on large volumes of data measured from sensors.	N_5
Express the unique identity of the seafarer. This is to capture the unique personal and job-related characteristics of the person, such as gender, height, susceptibility to seasickness, job description or seafaring experience. This implies a one-to-one digital twin-to-person relationship.	N_6
Facilitate digital twin-to-digital twin communication.	N_7
Perform safety checks or alerts when unsafe conditions are a reality, for example in the case of man overboard.	N_8
Augment the ability of the seafarer to perform activities, for example by providing instructions on how to do something, where fellow seafarers are or directions on where to go.	N_9
Present relevant information to the seafarer in a practical manner and facilitate interaction, such as responses to an instruction or indication of task progress, from the seafarer to the digital twin.	N_10
Capture the relevant state and behaviour relating to each person on board in close to real-time.	N_11
Store data and information in a sustainable and organised manner.	N_12
Facilitate capturing and managing future events and planned activities, such as a schedule, for the seafarer's personal and public use.	N_13

5.2 FUNCTIONAL FLOW DECOMPOSITION

A functional analysis was completed as a method of examining what actions a human digital twin solution would be required to perform as a system. The highest level of functional flow considered, the system-level, is shown in Figure 2. These functions can be used to break the system into subsystems with specific requirements to fulfil, however, this is often achieved for a context-specific application. The functional flow discussion will be directed towards a human digital twin for monitoring and managing human comfort, relevant to the use case described in Section 6.

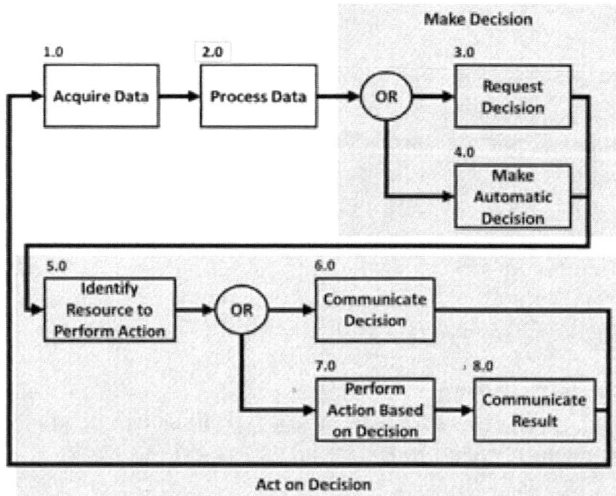

Figure 2: Functional Flow Block Diagram for a Human Digital Twin Solution

The functional block 1.0 in Figure 2 relates to obtaining sensor observations, or data from various sources, with which the state and behaviour of seafarers can be gathered and estimated. For example, data that is acquired with regards to human comfort could be accelerometer readings, which measure the structural vibration of the ship, or seafarer location readings of where the seafarer is on board.

Processing the data captured, considered at functional block 2.0 in Figure 2, could comprise of using models or standards to quantify human factors, such as using ISO 2631-1 [32] to quantify human comfort due to whole-body vibration. It is important that at this stage the human digital twin is updated, based on the most up-to-date data acquired, to represent the state and behaviour of the seafarer in near real-time.

The functions to follow, blocks 3.0 to 8.0 in Figure 2, exist largely to facilitate decision-making. It is important that the processed data from functional block 2.0 can be used to make decisions and allow for the extraction of actionable insight from the information provided from sensor readings in near real-time. In order to make a decision, a decision-maker is required. Either the human digital twin should be able to identify the appropriate decision-maker, through unique information and knowledge potential about each person stored in their digital twin that is visible to peer digital twins, or the seafarer should suggest who to ask to make a decision. The seafarer may want to ask other people face-to-face or by means of asking other seafarers through twin-to-twin communication. Thus, the digital twin implementation should have an infrastructure that facilitates communicating with peers, twin-to-twin.

In the case of making "automatic" decisions, seen in block 4.0 in Figure 2, the human digital twin is the decision-maker. An example decision may be an automated determination of the seafarer's comfort, based on thresholds. The human digital twin should also be able to learn from previous simple decisions made by the seafarer. By no means does this imply that humans will be subservient to the digital twins; rather, trivial or recurring decisions, that may lend themselves to frustration of the seafarer, are logged. This may also incorporate gaining permission from the seafarer at least the first time an "automatic" decision is made.

Similar to requiring a decision-maker when a decision is to be made, an action-performer is required when an action is to be performed. Once a decision has been made, functional blocks 5.0 to 8.0 represent acting on the decision, by means of performing an action as a response to the decision. Again, similar to making a decision, either the human digital twin or a seafarer can be the action-performer. An example scenario where an "automatic" decision and action may be useful is if the seafarer is unexpectedly busy securing equipment that shook loose during a storm and cannot request lunch be kept for them. The human digital twin may recognise this through knowledge that the seafarer had lunch scheduled, but is running late and may miss lunch. The human digital twin can decide on the most suitable lunch for the seafarer, based on previous lunch preferences, and order lunch to be kept. Once an action such as this is completed, it is important to note that although the functional flow is considered to numerically end at block 8.0, the last blocks flow to the first block again, indicating the system strives to achieve its missions on a continuous basis, not just once-off.

6. ILLUSTRATIVE USE CASE: HUMAN COMFORT MONITORING AND MANAGEMENT

In order to realise a human digital twin solution, as shown in Figure 1, for monitoring and managing the comfort of seafarers on a seafaring vessel, the communication between the physical and digital twins, as well as the digital representation of the seafarer (the digital twin itself), needs to be developed. Communication from the seafarer to their digital twin would likely be acquiring data from sensors, integrating the data from various sources and processing the data into information. An example of the embodiment of this is presented in Figure 3. The human comfort information would be presented to perhaps navigational officers in the bridge in an insightful manner to monitor human comfort and assist with decision-making towards managing human comfort. The communication to the seafarer from the digital twin could be, for example, via a smart watch the seafarer is wearing.

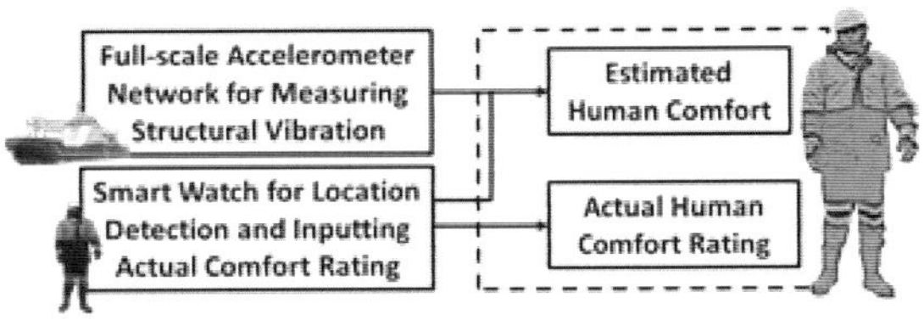

Figure 3: Example Embodiment of the Communication from the Physical World to the Digital World

6.1 MODELS OF HUMAN COMFORT

Shipping provides a complex, dynamic environment in which seafarers are exposed to various vibration stimuli from machinery on board, as well as the vessel's motion through the water. The human body is equally complex and dynamic, resulting in no simple or easily predictable response to vibratory stimulation [33]. Human comfort due to whole-body vibration, such as the structural vibration seafarers experience on a ship travelling through open water and ice, can be estimated and assessed by methods presented in ISO 2631-1 [32].

The ISO 2631-1 methodology evaluates whole-body vibration comfort based on frequency weighted root mean squared (RMS) acceleration values. The whole-body vibration exposure frequency range, in which human comfort is influenced, is of interest between 0.5 and 80 Hz in all axes of vibration. The associated vibration of each axis can be measured using accelerometers proximate to the vibrating surface interface on which a seafarer's body weight is supported. The ISO 2631-1 methodology hinges on a basicentric axis system of the human body, not the same axis system of the vessel according to which structural vibration accelerometers are likely orientated, which means that the orientation of the seafarer with respect to the vessel is an important factor to consider. The severity of the whole-body vibration perceived is also dependent on the posture of the seafarer, whether standing, seated or recumbent. An example methodology for calculating the whole-body vibration comfort of a bow-facing, standing seafarer is shown in Figure 4.

First, acceleration measurements are frequency weighted in the time domain. The root mean square acceleration is calculated for each vibration axis to determine the component ride value. Whole-body vibration comfort is related to the overall ride value, which is determined by accruing the component ride values using root sum of squares (r.s.s.) of the RMS values. The overall ride value is then classified in terms of comfort by considering the likely human responses which are provided in the standard and are correlated with likely responses of persons to vibration in public transport.

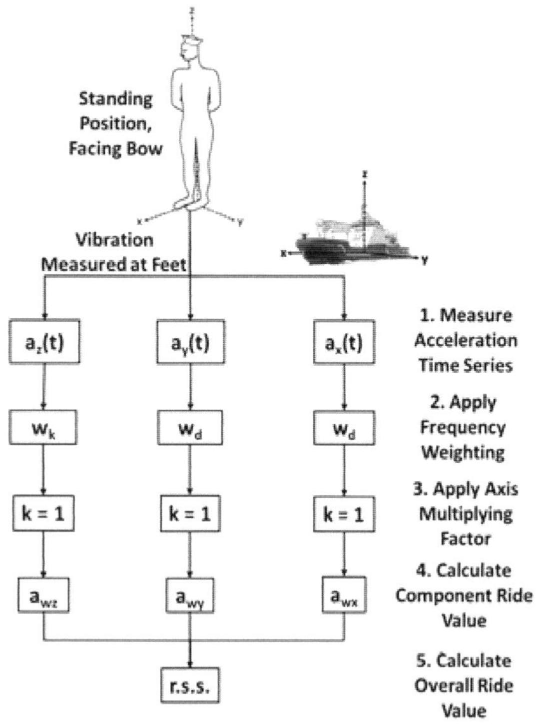

Figure 4: Method for Calculating the Whole-body Vibration Comfort of a Bow-facing, Standing Seafarer

Furthermore, the actual comfort of seafarers can be monitored by using the seafarer as a living sensor to rate their perceived comfort or discomfort through a convenient interface such as a smart watch or tablet. The perceived comfort of the seafarers captured in conjunction with the estimated comfort information can be used to develop personalised comfort models for seafarers over time. Such data could serve to improve the power and applicability of data-driven predictive models.

6.2 HUMAN COMFORT MONITORING DASHBOARD

An example of an envisioned interface, or monitor, which could be presented on a screen in the bridge for navigating officers to use as a dashboard for human comfort monitoring, is shown in Figure 5. This monitor displays various aspects of what could be presented to navigating officers, but would ideally be customised according to the vessel operator requirements.

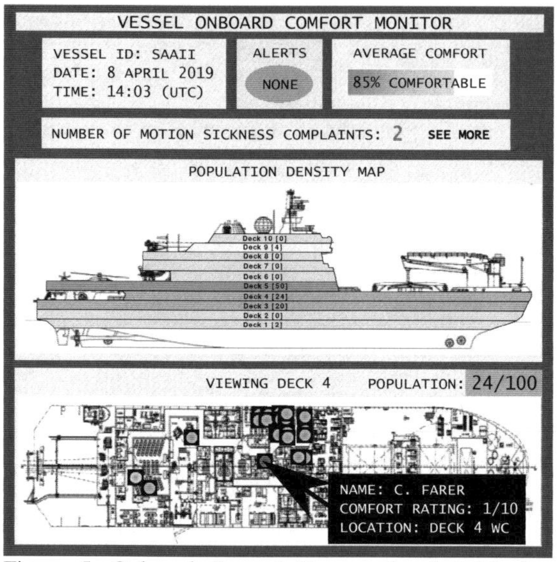

Figure 5: Onboard General Human Comfort Monitor Prototype

The monitor could enhance the officer's state awareness of seafarers by providing insight into the general comfort state of passengers shown via a bar graph indicator, as seen in Figure 5, that is supplemented with text and a green colour to indicate that the average seafarer is about 85 % comfortable. The monitor could similarly augment the officer's spatial awareness of seafarers on board by displaying the population density map for all decks. Furthermore, the situational awareness could be deepened by, for example, being able to view the location of a passenger that lodged a motion sickness complaint. By viewing the deck that the seafarer of interest is on, the said passenger can be indicated, as in Figure 5, with a red coloured dot (symbolising severe discomfort) to be situated at the deck 4 water closet.

To gain further insight into the situation, the officer could click on the likely motion sick passenger to view their personal comfort monitor - an example of which is shown in Figure 6. Upon viewing the personalised human comfort dashboard for the passenger that complained of motion sickness, the officer's situational awareness is deepened by seeing that this passenger's comfort started declining around the time of a storm that started (not indicated on the dashboard here). Furthermore, the passenger is seen to recently have been in the dining room, which should explain why their current location at the water closet coincides with a motion sickness complaint. A message could be sent to the passenger to recommend taking motion sickness medication and laying down for a while before continuing with their planned activities.

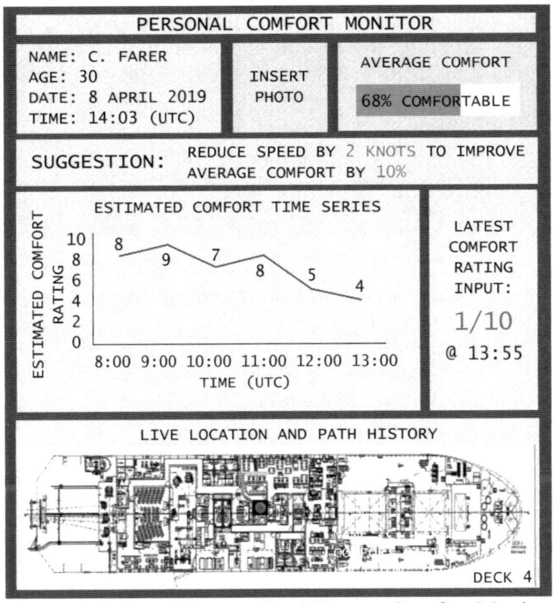

Figure 6: Onboard Personalised Human Comfort Monitor Prototype

7. CONCLUSIONS AND FUTURE WORK

Maritime 4.0 is transforming the maritime industry, including the working and living environment for seafarers. People are still integral to many vessel operations and, especially in the case of the passenger sector, critical to success. Human-centric development within the Maritime 4.0 movement, that is often machine and system focussed, is important for the future of the maritime industry. In this paper, human-related challenges in seafaring vessels were identified and discussed. A human digital twin solution which offers to facilitate the integration of seafarers into Maritime 4.0 environments, resulting in the development of a smart, connected Mariner 4.0, is proposed. This includes the current conceptual development of Mariner 4.0 on the SA Agulhas II. The expected value of a human digital twin solution is discussed, revealing a human-centric digital platform that, along with the link between the cyber and physical worlds, offers the potential to facilitate important communication, transfer of knowledge (or training), decision-making and data management. The result of such a solution is envisioned to lead to improved safety and efficiency of operations on board.

Though Maritime 4.0 offers potential solutions to current and future human-related challenges, human-centric research efforts and developments within Maritime 4.0 are still essential. Planned future work relates to the further development of the Mariner 4.0, including the development and implementation of a human digital twin architecture with specific focus on monitoring and managing the well-being of passengers and crew members on board the SA Agulhas II.

8. ACKNOWLEDGEMENTS

The financial assistance of the National Research Foundation (NRF), through the South African National Antarctic Programme (SANAP Grant No.110737), towards this research is hereby acknowledged. Opinions expressed and conclusions arrived at are those of the authors and are not necessarily to be attributed to the NRF.

9. REFERENCES

1. Romero, D., Bernus, P., Noran, O., Stahre, J., Fast-Berglund, A., 'The Operator 4.0: Human Cyber-physical Systems And Adaptive Automation towards Human-automation Symbiosis Work Systems', *Proceedings of the International Federation for Information Processing on Advances in Production Management Systems*, Iguassu Falls, Brazil, pp. 677-686, 2016.

2. Sullivan, B. P., Desai, S., Sole, J., Rossi, M., Ramundo, L., Terzi, S., 'Maritime 4.0 – Opportunities in Digitalisation and Advanced Manufacturing for Vessel Development', *Proceedings of the International Conference on Industry 4.0 and Smart Manufacturing*, Rende, Italy, pp. 1-8, 2019.

3. Heilig, L., Lalla-Ruiz, E, Voβ, S., 'Digital Transformation in Maritime Ports: Analysis and a Game Theoretic Framework', *Netnomics*, Volume 18, pp. 227-254, 2017.

4. Rojko, A., 'Industry 4.0 Concept: Background and Overview', *International Journal of Interactive Mobile Technologies*, Volume 11, Issue 5, pp. 77-90, 2017.

5. Sparrow, D., Kruger, K., Basson, A., 'Human Digital Twin for Integrating Human Workers in Industry 4.0', *Proceedings of the International Conference on Competitive Manufacturing*, Stellenbosch, South Africa, pp. 259-265, 2018.

6. Tzafestas, S. G., 'Concerning Human-Automation Symbiosis in the Society and the Nature', *International Journal of Factory Automation, Robotics and Soft Computing*, Issue 3, pp. 16-24, 2006.

7. Cicek, K., Akyuz, E., Celik, M., 'Future Skills Requirements Analysis in Maritime Industry', *Proceedings of the World Conference on Technology, Innovation and Entrepreneurship*, Istanbul, Turkey, pp. 270-274, 2019.

8. Lützhöft, M., Hynnekleiv, A, Earthy, J. V., Petersen, E. S., 'Human-centred Maritime

Autonomy – An Ethnography of the Future', *Journal of Physics: Conference Series*, Volume 1357, 2019.

9. Erikstad, S. O., 'Merging Physics, Big Data Analytics and Simulation for the Next-generation Digital Twins', *Proceedings of the Conference on High-performance Marine Vehicles*, Zewenwacht, South Africa, pp. 140-150, 2017.

10. Bruynseels, K., de Sio, F. S., van den Hoven, J., 'Digital Twins in Health Care: Ethical Implications of an Emerging Engineering Paradigm', *Frontiers in Genetics*, Volume 9, Article 31, pp. 1-11, 2018.

11. Bekker, A., 'Exploring the Blue Skies Potential of Digital Twin Technology for a Polar Supply and Research Vessel', *Marine Design XIII*, Volume 1, pp. 135-146, 2018.

12. Casareale, C., Bernardini, G., Bartolucci, A., Marincioni, F., D'Orazio, M., 'Cruise Ships Like Buildings: Wayfinding Solutions to Improve Emergency Evacuation', *Building Simulation*, Volume 10, Issue 6, pp. 989-1003, 2017.

13. Elnabawybahriz, M. N., Hassan, M. H. N., 'The Impact of Low Efficient Evacuation Plan During Costa Concordia Accident', *International Journal of Mechanical Engineering*, Volume 5, Issue 1, pp. 45-56, 2016.

14. Kwee-Meier, S. T., Bützler, J. E., Schlick, C., 'Development and Validation of a Technology Acceptance Model for Safety-enhancing, Wearable Locating Systems', *Behaviour and Information Technology*, Volume 35, Issue 5, pp. 394-409, 2016.

15. ScanReach, 'In:Range', Retrieved at: https://www.scanreach.com/solutions/#solutions-inrange, Accessed: 15 January 2020, 2019.

16. Swedberg, C., 'Wi-Fi System Employed by DHS Tracks People Location and Traffic', *RFID journal April*, Retrieved at: https://www.rfidjournal.com/wi-fi-system-employed-by-dhs-tracks-people-location-and-traffic, Accessed: 5 June 2020, 2020.

17. Ligthelm, F., SA Agulhas II mini-symposium presentation, 29 November 2019, Stellenbosch, South Africa.

18. Nautilus International, 'An Investigation into Connectivity at Sea', *Connectivity at Sea White Paper,* June, Retrieved at: https://www.nautilusint.org/en/news-insight/resources/nautilus-reports/connectivity-at-sea-whitepaper/, Accessed: 15 January 2020, 2017.

19. Lim, K., 'Maritime Communications – Safeguarding the Spectrum for Maritime Services', *ITU News Magazine*, Volume 5, pp. 68-71, 2019.

20. Hand, M., 'Isolation and the Challenges of Crew Internet', *Seatrade Maritime News*, November, Retrieved at: https://www.seatrade-maritime.com/americas/isolation-and-challenges-crew-internet, Accessed: 5 June 2020, 2018.

21. Christos, G., 'Key Challenges Toward the Way from Smart Shipping to Internet of Ships', *Safety4sea*, June, Retrieved at: https://safety4sea.com/key-challenges-toward-the-way-from-smart-shipping-to-internet-of-ships/, Accessed: 15 January 2020, 2018.

22. Lázaro, F., Raulefs, R., Wang, W., Clazzer, F., Plass, S., 'Very High Frequency Data Exchange System (VDES): An Enabling Technology for Maritime Communications', *Council of European Aerospace Societies Space Journal*, Volume 11, pp. 55-63, 2018.

23. Erikstad, S. O., 'Designing Ship Digital Services', *Proceedings of the Conference on Computer and Information Technology Applications in the Maritime Industries*, Tullamore, Ireland, pp. 458-469, 2019.

24. Nowak, J., Stakkeland, M., 'Implementation of Data Driven, Iterative Approach to Building Digital Twins', *Proceedings of the Conference on Computer and Information Technology Applications in the Maritime Industries*, Tullamore, Ireland, pp. 359-373, 2019.

25. Özsever, B., Tavacıoğlu, L., 'Analysing the Effects of Working Period on Psychophysiological States of Seafarers', *International Maritime Health*, Volume 69, Issue 2, pp. 84-93, 2018.

26. European Community Shipowner's Associations, 'The Future of the Modern Seafarers Profession', *The Maritime Executive*, June, Retrieved at: https://www.maritime-executive.com/editorials/the-future-of-the-modern-seafarers-profession, Accessed: 24 January 2020, 2018.

27. Kritzinger, W., Karner, M., Traar, G., Henjes, J., Sihn, W., 'Digital Twin in Manufacturing: A

Categorical Literature Review and Classification', *Proceedings of the 16th International Federation of Automatic Control Symposium on Information Control Problems in Manufacturing*, Bergamo, Italy, pp. 1016-1022, 2018.

28. Munir, S., Stankovic, J. A., Liang, C. M., Lin, S., 'Cyber-physical System Challenges for Human-in-the-loop Control', *Proceedings of the 8th International Workshop on Feedback Computing*, San Jose, California, 2013.

29. Romero, D., Stahre, J., Wuest, T., Noran, O., Bernus, P., Fast-Berglund, A., Gorecky, D., 'Towards an Operator 4.0 Typology: A Human-Centric Perspective on the Fourth Industrial Revolution Technologies', *Proceedings of the 46th International Conference on Computers and Industrial Engineering*, Tianjin, China, pp. 1-11, 2016.

30. Sparrow, D. E., Kruger, K., Basson, A. H., 'An Architecture for the Integration of Human Workers into an Industry 4.0 Environment', *Submitted to the International Journal of Production Research*, 2020.

31. Taylor, N., Human, C., Kruger, K., Bekker, A., Basson, A., 'Comparison of Digital Twin Development in Manufacturing and Maritime Domains', Service Oriented, Holonic and Multi-Agent Manufacturing Systems for Industry of the Future, Cham: Springer, *Studies in Computational Intelligence*, Volume 853, pp. 158-170, 2020.

32. International Organisation for Standardisation, 'Mechanical Vibration and Shock – Evaluation of Human Exposure to Whole-body Vibration – Part 1: General Requirements', *ISO 2631-1*, Geneva: International Organisation for Standardisation, 1997.

33. Griffin, M. J., 'Handbook of human vibration', *1st paperback edition*, London: Academic Press, 1996.

10. AUTHORS BIOGRAPHY

Nicole Taylor is currently a Master of Engineering research candidate at Stellenbosch University in South Africa. Her thesis project is entitled *"Implementation of a Human Digital Twin for Monitoring Motion Sickness of Passengers and Crew on a Polar Research and Supply Vessel"*.

Anriëtte Bekker holds the current position of Associate Professor at the Department of Mechanical and Mechatronic Engineering Stellenbosch University. Here, she directs the Sound and Vibration Research Group, leading and supporting final year, Master and Doctoral engineering candidate research.

Karel Kruger holds the current position of Senior Lecturer at the Department of Mechanical and Mechatronic Engineering Stellenbosch University. He is also the co-director of the Mechatronics, Automation and Design Research Group, supervising final year, Master and Doctoral engineering candidate research projects.

FEM ANALYSIS OF AN ICE CLASS SHIP'S BULBOUS BOW

B. D'Amato, G. Soriano, A. Ambrosio, University of Naples Federico II, Italy

SUMMARY

Ice Class Ships are supposed to be designed and built in a different way other than conventional vessels. In this article will be approached a FEM analysis of an ice class ship's bulbous bow. Ice Class is an additional class notation that implies several measures, since these vessels are often subject to special loads due to ice, it is necessary to ensure optimal ship performances even on ice.
A Special attention has to be given to the structures and geometry of the bow sections in fact, the bow has to be designed in order to define the best shape to be used, so to minimize the hull's resistance.
A FEM analysis is the most reasonable approach to the bow's design, as this structure is complex to analyse and to be studied as an elementary geometry. The FEA allows a more accurate design with a possible reduction in weight, complying with the rules.
The structure analysed has been defined from the collision bulkhead, where the constraints are considered as joint to the rest of the hull. The article pursues on a static analysis: hydrostatic forces have been applied to the bow, in order to have as results consequent shifts.
The case study is a refitting of an Oil-Tanker in an Ice Class IC Research Vessel, has been analysed with the FEM Software PASEF, developed in the University of Naples Federico II.
Results will show that the optimized bow structure is strong enough to stand the different loads.

NOMENCLATURE

β	Stress concentration factor (-)
γ	Security factor (-)
η	Correction coefficient (-)
σ	Stress (MPa)
σ_1	Stress component (MPa)
σ_2	Stress component (MPa)
σ_3	Stress component (MPa)
ρ	Density of water (kg m^{-3})
τ_{yz}	Shear component (MPa)
τ_{xy}	Shear component (MPa)
τ_{xz}	Shear component (MPa)
σ_{amm}	Maximum admissible stress (MPa)
σ_{eq}	Equivalent stress (MPa)
A	Area (m^2)
c_a	Probability that the hull is subjected to the same ice pressure at the same time (-)
c_d	Influence of the size of the ship and its propulsion system (-)
c_l	Probability that the design pressure occurs in that part of the hull (-)
C_{st}	Coefficient of collaboration (-)
F	Pressure force (N)
F_1	Correction coefficient (-)
g	Gravity acceleration (m s^{-2})
H_R	Ridge thickness (mm)
int (	Integral
k	Correction coefficient (-)
p	Pressure exerted by the design ice equal to $c_d\, c_l\, p_0$ (MPa)
p_0	5,6 (MPa)
Ppl	Pressure exerted by the ice on the shell (MPa)
P_S	Hydrostatic pressure (N m^{-2})
R_B	Breaking resistance (N)
R_{eH}	Minimum yield stress of the material (MPa)
R_F	Friction (N)
R_i	Resistance due to ice (N)
R_{iTOT}	Total resistance due to ice (N)
R_{ow}	Resistance in open water (N)
R_R	Ridge resistance (N)
R_{RB}	Ridge breaking resistance (N)
R_S	Resistance of the submerged part (N)
$R_{viscous}$	Resistance due to viscosity (N)
R_{wave}	Resistance due to waves (N)
$R_{wave\text{-}viscous}$	Resistance that considers the interaction between R_{wave} and $R_{viscous}$ (N)
s	Distance between ordinary stiffeners (m)
t_c	Extra thickness due to corrosion and abrasion (2 mm)
T	Draft (m)
z	Considered point within to the base line (m)

1. INTRODUCTION

1.1 ICE CLASS SHIP & ADDITTIONAL CLASS NOTATION

An Ice Class Vessel is supposed to be built in a different way other than conventional vessels: in fact, not all ships are built according to the ice class requirements, this is an additional class notation adopted from a register.
Ships classified with Ice Class will probably increase because of global warming, since the shrinking of the Arctic ice pack in the future, may open new sea routes, as for example the Northeast and Northwest passages.
Ice Strengthened Ships are classified as vessels able to sail in ice, but with the aim of an Icebreaker in ice conditions; these ships usually have a bulbous bow.
Several registers have rules for Ice Class classification, for example the RINA 2017 [1] defines the following additional class, the notations are assigned in accordance with Pt A, Ch 1, Sec 2, [6.10] to ships strengthened for navigation in ice and complying with the relevant requirements of the Pt F, Ch 9, Sec 1:

- ICE CLASS IA SUPER
- ICE CLASS IA
- ICE CLASS IB
- ICE CLASS IC
- ICE CLASS ID.

The Ice Class IA Super is built for navigation without the additional support of an Ice Breaking Ship, while the Classes from IA to ID are built with decreasing ice strengthened structures.
The RINA regulations for Ice Class vessels will be applied in this project.
Other requirements concern:

- main engines, that require additional nominal power,
- additional thickness for the propeller blades,
- a higher shell height, for the bow,
- welding between the hull and the bilge keels need to be reinforced,
- a maximum speed for every Ice Class.

Historically the general arrangement of ships enabled to sail in ice regions changed in a few ways. The most remarkable change concerns the accommodation: from late 1970's they are not placed in the hull in order to increase the height of the bridge to have a better visibility and, in part, to avoid the vibration and noise caused by ice creep.
In this article the main loads that will occur on an Ice Class ship will be analysed, both from the ice and water, and then the consequent structural stresses. An application with FEA will be carried out applying and explaining the theory and the best way to proceed, applied to a case study.

1.2 THE BOW

The bow is the forward part of the ship, the first to cut the waves, and its shape is essential to define the entire ship resistance and, consequently, the power installed and specific consumptions. [2] As in case of displacing ships with fuller shapes the Wave breaking resistance has significant components in the total resistance of the vessel, while ships with a different shape, slender and more curved, have less wave breaking resistance.
So, there are several factors that influence the design of the bow, it depends on the cruise speed, the route and the service of the vessel.
The most common shapes are divided in three main categories [3]:

- Conventional bow, without a bulb,
- Special bows,
- Bulbous bow.

A conventional or normal bow is an evolution of the vertical bow, which was so popular in the first 1900 liners. Nowadays an example is given by a vertical bow with a straight edge, that is called *plumb bow*, and these bows have the maximum waterline, that allows for a greater hull speed.
Special bows are intended as unconventional designed, that may appear out of the ordinary like X-Bows, well known as inverted bow, that have the role of improve the seakeeping in rough seas, a reason why this type of bow has been often adopted in the Explorer Yachts and in a new concept of cruising vessels.[4]
In the 90s the bulbous bow was the most common bow shape to be adopted in the ship design, the most common reason of its widespread use is the reduction of wave resistance and the consequent lower fuel consumptions that, especially with high cruise speed, has an high impact on the vessel's operational costs. [5]

1.3 BULBOUS BOW

A ship has three components of the hull's resistance, ideally [5]:

- R_{wave}
- $R_{viscous}$
- $R_{wave\text{-}viscous}$

with the following assumptions taken:

- the vessel surges in a viscous fluid with free indeformate surface,
- other components, the apparent resistances are neglected.

These components have been analysed experimentally and have been associated with the dissipation of mechanical energy in the wake of the vessel, and to the energy flow of the wave system.
The component due to the waves can be easily modified by changing the shape of the first part of the ship to enter the waves: the bow.
A ship with a normal bow sails in still water generating its own wave, as shown in cyan in the picture, such a wave is a synonym of dissipated energy, caused by the moving mass of the ship transferred to the water, so it is a resistance [5].

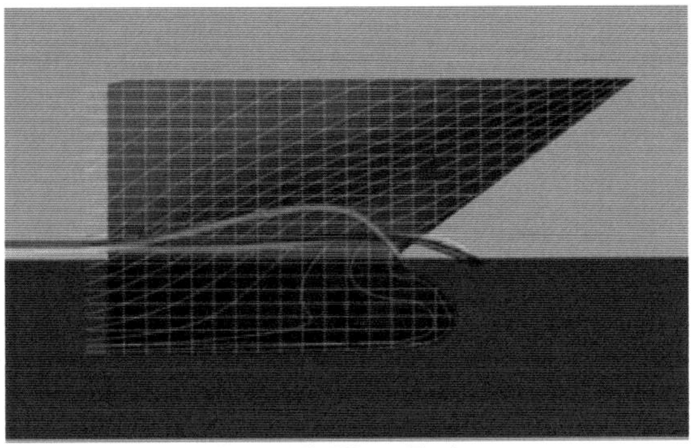

Figure 1: Wave Trains

The idea of introducing a bulb was born in the late 1970s where the first studies carried out experiments on a sphere that generates its own wave train (in the picture is shown in red). These studies demonstrated that a basin model with a sphere placed ahead of the bow, with specific dimensions and distances from the hull has a reduced wave resistance.

In fact, modifying the distance and dimensions of the sphere it has been demonstrated that a new wave train can be generated, in phase opposition with the waves generated by the hull, as shown in green in the picture above. The resulting wave is minor in height than the hull's one without the bulb, this means that the ship dissipates less energy in moving into the water, so the wave resistance is reduced.

In Ice conditions adopting a bulbous bow is recommended not only for the advantages coming from the drag reductions due to the shape itself, but also the bulb allows broken ice to move smoothly along the hull, avoiding dangerous collisions. The immersed side of the ice, that has less friction coefficient than the emerged part, reduces the total resistance of the ship.

So, adopting a bulb in a design process of an Ice Class ship is a proper choice, since it is navigating with an ice breaker vessel and doesn't need to break the ice.

2. PERFORMANCE ON ICE

The ship's performance in ice is evaluated by its ability to break ice and maneuver in ice. These capabilities were considered in concept design and defined in the contract's specifications. This performance depends on the vessel's mission profile. The skill of breaking ice is measured in uniform ice conditions in function of operability speed at which certain ice thickness can be broken.

The ship's speed in ice is determined by the ice resistance depending on the thrust provided by the propulsion, and the hull shape and main dimensions as well as ice properties. The manoeuvring performance is similarly determined by the transverse forces provided by the rudder(s)/azimuthing thrusters and the resisting forces mainly due to ice [6]. Therefore, performance in ice is influenced by drag forces and propulsive forces and these can be improved by the hull shape minimizing drag forces and propulsion design maximizing propulsive forces [6].

This article will focus mainly on the evaluation of the added resistance due to ice in the water, since the ship in question is an ice class ship and not an icebreaker. In fact, for ships that are not icebreakers, the ability to break ice is not evaluated.

2.1 ICE RESISTANCE

Ice resistance refers to all the longitudinal components of forces due to ice acting on the ship over time, so more than a static problem it results in a dynamic one. These ice forces are divided into categories of different origin [6]:

- Breaking forces;
- Submergence forces;
- Sliding forces.

In different ice conditions the relative importance of these components varies; in level ice the breaking component is usually the largest but in brash ice or in smaller ice floes the other two components become more important. The breaking force is related to the breaking of the ice i.e. to crushing, bending and turning the ice. In this case study, the breaking force isn't considered. Submergence is related to pushing ice down along the ship hull whereas the sliding forces include frictional forces [6]. Usually the velocity dependency of the ice resistance is attributed to the last component.

Ice resistance in level ice is at the basis of all other ice resistance formulations, this is investigated first. If a test is made in uniform level ice where the ship power is kept constant, the ship eventually reaches a constant speed. The total resistance in ice, R_{iTOT}, can be assumed to be equal to the propeller thrust (with the so-called thrust deduction deducted). If the power is decreased, a new, lower speed is reached, and a new ice resistance point can be obtained. The total resistance in ice is assumed to be the sum of the pure ice resistance R_i and open water resistance R_{ow}

$R_{iTOT} = R_i + R_{ow}$

This assumption can be made as first approximation due to the nature of the phenomena, since it is not possible to describe it in an exact solution [6]. The total resistance in ice and the open water resistance can be determined experimentally in model tests and then the pure ice resistance can be determined by subtraction. The ice resistance is further divided into components mentioned above, thus the ice resistance is

$R_i=R_B+R_S+R_F$

where the components are the breaking, submergence and friction component, respectively. Most methods used to calculate the ice resistance are based on regression on full scale and model scale data. The regression assumes the ice resistance to be linear with ship speed and to consist of these three components. In this case $R_B=0$, therefore

$R_i=R_S+R_F$

Thus, the calculation methods for ice resistance are at best semi-empirical, and these methods should be used cautiously, especially outside the range of validity. The calculation methods to determine the ice resistance should be used only in the conceptual design phase as these methods cannot account for the details of the hull shape.

When the design proceeds, ice model tests should be carried out to finalize the hull shape.
The ice resistance in broken ice can be determined similarly as the ice resistance in level ice. The only exception is that the breaking component is different but here the breaking force is not considered.
Ice resistance in ridges is dealt with similar methods as in brash ice. The major difference is, however, that as brash ice resistance depends on brash ice thickness H, the ridge resistance depends similarly on the ridge thickness, which is different at each location along the hull. Thus, the resistance from a ship length segment Δx at the location x is:
$\Delta R = R_R(H_R(x))\ \Delta x$
where $R_R(H_R(x))$ is the ridge resistance in ridge thickness H_R per unit ship length [6]. Here the ship speed is not mentioned as the ridge resistance is commonly treated as speed independent and the speed dependency is allocated to the open water resistance. The total resistance in a ridge is thus:
$R_{R,TOT} = R_{R,B}(H_R(x_{bow})) + int(R_R(H_R(x))dx) + R_{ow}(v)$.
The length of the ship's parallel midbody is denoted as L_{PAR}. The integration is for the length of the parallel midbody. The speed dependency in open water resistance is emphasized. In this work, the $R_{R,B}$ has been neglected, so the previous one becomes:
$R_{R,TOT} = int(R_R(H_R(x))dx) + R_{ow}(v)$.
As the ship is moving, location of the ridge relative to the ship is changing (x = x(t)) and as the ridge is not of uniform thickness, ridge resistance is changing constantly with time. It is consequently more suitable to speak of the energy required to penetrate a certain size of ridge.

2.2 PERFORMANCE IN ICE

Measures by which the ship performance in ice is described can be seen from the functional specification described above. These measures can include [7]:

- Speed(s) achieved in certain level ice thickness;
- Penetration of certain size ridges with a stated impact speed;
- Ship turns 180°in less than a certain time with a specific ice thickness.

Ship design proceeds so that at an early design phase some analytical methods are applied to determine the level ice performance, ridge penetration and brash ice performance. It is difficult to find a comprehensive presentation of methods to be used. The methods presented in references are at best rudimentary and as there are no analytical methods that can be applied in determining the manoeuvring performance, ice model tests are necessary at the end of conceptual design phase to verify the design. Turning performance in ice is measured by the diameter of the turning circle (divided by the ship length). It is Important to perform certain manoeuvres in the shortest time possible. A good manoeuvring capability can be achieved by a proper hull form design, having a large rate of turn of the rudders/azimuthing thrusters and providing a large transversal component of forces. The analysis of turning performance on ice wasn't measured in the paper. Ship performance in ridges is not measured by ridge resistance or any speed reached in ridges as commonly the largest ridges cause a resistance, even defined as an average value, that is so large that the delivered thrust cannot overcome it. Ridges are penetrated by consuming the kinetic energy of the vessel; thus, the correct parameter for ridge capability is the energy required to penetrate ice ridges. This depends on ship displacement, ship main dimensions and on bow shape.

2.3 HULL SHAPE DESIGN

The hull shape design of ice class ships aims at:

- Minimizing the ice resistance by selecting optimal beam and bow shape;
- Ensuring good operational (manoeuvring) characteristics;
- Enabling the ship to go astern as much and as well as the operational description requires;
- Ensuring a proper undisturbed operation of the propeller(s) by minimizing the amount of ice impacting on the propeller(s).

The most important parameters for ice resistance are the beam B and the stem angle φ [6]. A large beam causes more resistance and thus narrower ships with a large L/B ratio (especially if there is a draught restriction) is the result.
This article wants to focus mainly on Minimizing the ice resistance by selecting the optimal bow shape.
Bow shape should be such that it allows the ice floes to float towards the surface before getting under the bottom. At present there is no alternative to ice model testing as analytical methods are not so advanced that they could predict the effects which can be deemed local. Here only one effect is mentioned. This one concerns the bulbous bow. It has often been stated, and this has been reflected in some classification society ice rules, that bulbous bow is not good in an ice going ship. This is because the bulb itself does not break ice very well and also because there is one frame that is vertical at the bow, if the ship has a bulbous bow (at the bulb ice is bent up and at the shoulders down; thus there is a vertical frame in between).
But present experience shows that most merchant ships don't need to break ice as they either sail in broken channels or follow an icebreaker. Thus, ice strengthened ships often have bulbous bow which is not a handicap in broken ice. The reason for this is that broken ice is displaced around the hull in a way that resembles the hydrodynamic flow. Only in ice going ships and icebreakers which must break ice themselves the bulbous bow is not appropriate. By shaping the bulbous bow for ice, much of the additional ice resistance can be avoided.
In this case study, the design of the hull shape has been already defined by the designers, but only the stresses due to the pre-existing shapes have been evaluated.

3. STRUCTURAL MODEL

3.1 FEM THEORY & MODELING - FROM LOAD TO ELEMENTARY MODEL

The theory behind the Finite Elements Method develops in several steps and allows a complete analysis of all the stresses occurring in the structure, at different loads.
The FEM is a numerical method that, in order to solve a problem, divides the structure into several smaller elements, pieces of the structure itself, and analyses every element at one time. The question is to find approximate solutions of problems described by partial differential equations. For every element is possible to define in input the loads and the constraints and then, through algebraic equations, shifts can be determined.
Since the number of equations may be remarkable and also their resolution may not be so easy, several softwares have been developed.
Depending on the matrix that defines the elementary piece of the structure, these are the main different elements [8]:

- plate
- plate with stresses in a plan-load
- beam

The difference between each one is the number of equations and solutions that define the matrix.
The division into smaller elements is based on the concept of discretizing a continuous system, and then every element can be analysed separately: in this way a mesh can be created, and it will be made of a finite number of points.
Mathematically, the formulation of a boundary value problem results into a system of algebraic equations that are simpler to solve than the partial differential equations, that describe the continuous system.
These elements differ not only for the number of equations, but analysing the problem through the stresses, an easy guideline can be given.
The plate is a structural element with the two main dimensions bigger than the thickness, with a bending behaviour that considers the deformations in perpendicular to the mean plane: the stresses are defined in three directions.
Another case of plate can be defined as a solid with the following properties:

- Thickness: one of the three dimensions of the plate is smaller than the other two.
- Flatness: The middle surface of the plate is a plane and is defined as the locus of points that halves the transversal fibers.

Thanks to these hypotheses, when the loads act on the borders in parallel to the middle-surface of the plate, the stresses are in a plan-load, and if the material is linearly elastic, the plate simplifies into a two-dimensional elastic problem.
The beam follows the Eulero-Bernoulli beam theory, a simplification of the theory of elasticity in a linear domain. It is an ease of the De Saint Venant problem, thanks to the hypothesis of conservation of plane sections, by which the sections are bounded to not move outside their plane, and once the rotation angle of the section within respect to the beam axis is known, is also possible to determine the dimensional shifts of every point of the section.
The element chosen to outline the planking of the hull is the plate that considers the stresses acting in three directions, since also shifts perpendicularly to the hull should be defined.

3.2 DEFINING THE CONSTRAINTS

A rigid body in space is characterized by 6 degrees of freedom (DOF) (three due to translations, and three to rotations) instead, in a plane problem a rigid body has 3 DOF [8]. The entire beam of the ship can be considered as a free rigid body with 6 DOF but considering a section of the ship is a completely different case.
In fact, the structure considered is not free in space anymore, it must be considered bounded with the rest of the ship.
A solution that fits also in softwares application is to consider the bow section bounded with constraints.
The constraints to be adopted are joints, between the bow section and the remaining of the hull, in order to restore the continuity with the structure.

3.3 LOADS ANALYSIS

The hull girder of an Ice Class vessel is subject to dynamic and static loads, as seen in the previous paragraphs, with a consequent response on the structure [5].
As an example, the static loads in still water have two components: buoyancy and gravity (or weights). The buoyancy is the resultant of the hydrostatic pressure distribution on the immersed surface of the ship. This pressure is a force that acts perpendicularly on the surface, in every point of the immersed hull. The buoyancy is the resultant force that acts perpendicularly to the waterplane area and directed upwards.

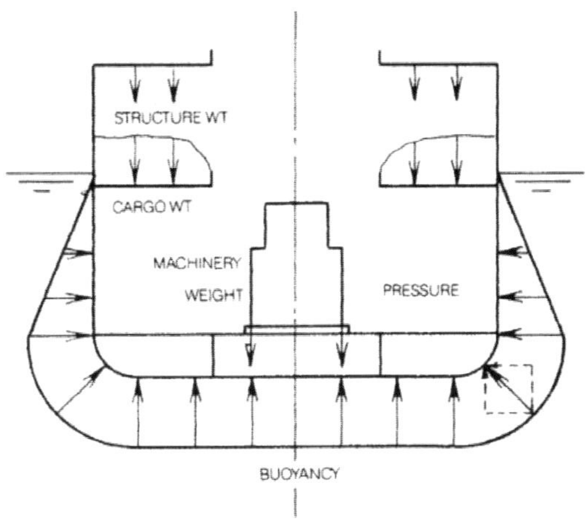

Figure 2: Buoyancy and Weights

Individual loads may have local and global structural effects. For example, the main engine induces high local

loads on a small surface, so the basement should be designed in order to distribute along the entire hull the loads due to the weight.
It should be noted that the geometry and the resulting evolution of the tensional and bending response patterns of the ship structures, are often divided into a scheme. The hull is divided depending on the specific part that is subject to a specific load, it is usually convenient to divide the structure and the associated response into three components, that are labeled as primary (global), secondary and tertiary (local).

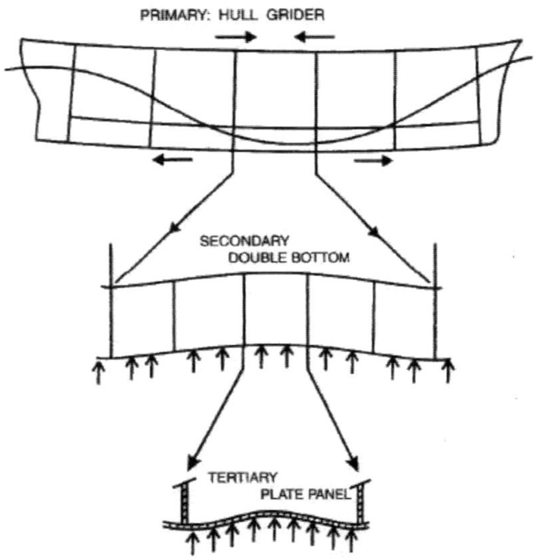

Figure 3: Loads Distribution

Other loads, that are not classified as static loads, as ice loads, have similar structural responses, so the ice loads follow the same pattern. In the application with the FEM software that will follow, the static loads have been considered acting perpendicularly to the hull in every point. The loads have been applied as a single force in the middle of every finite element (plate), in order to simulate the distributed pressure. The loads have been applied on the overall section, except from the points that are joint to the rest of the hull, since the software reads it as etched shifts.

4. A CASE STUDY

A case study has been required by the University as a project. Will follow an analysis on the previous ship and the need of a refitting operation, with a practical application through a FEM software that will study the bow section of an Ice Class Vessel.

4.1 REFITTING IN ICE CLASS IC

The ship to be refitted was originally a small Oil tanker and, after being sold to a research institute, the new mission profile is a research vessel, able to sail in ice conditions.
The university required a FEA in still water on the bulbous bow that has been installed, since the previous bow was a straight one. Static loads have been analysed, and the minimum thickness of the planking has been determined, based on the value that obtains as result zero shifts after applying the loads. Diaphragms have been considered and outlined as the element plate, setting their own thickness.

4.2 SOFTWARE

The work described in this article was developed using the software PASEF [9]. PASEF is the acronym for "Programma Analisi Strutturale ad Elementi Finiti" (Program for Finite Element Structural Analysis). The program was developed within a project of the structural engineering department of the Federico II University of Naples. It was made starting from the idea of Professor Russo Spena and Professor Sparacio in the 1980s. Its first use was in the field of civil engineering for the analysis of steel and reinforced concrete structures.
Today it has been implemented over time and is still used in the master's degree program in naval architecture and marine engineering during the course of *construction science 2* carried out by the Prof. Russo Spena. Every year dozens of students carry out university projects by making a structural analysis of a specific part of a naval structure using this software. The software is developed in the DOS environment.
It has a simple and sequential user interface that allows you to define and modify the structure previously defined in another modelling software, for example Rhinoceros.
By entering the various codes from 1 to 12 the program allows to define the structure by entering: number of nodes, number of rods, number of beams, number of plates, number of solids and the characteristics of the gravitational acceleration relative to the structure.
Subsequently it is possible to define the elastic and mechanical properties of the groups of elements and by which nodes the various elements are defined with their material and their mechanical and geometric properties. Once the elements have been described, the forces and displacements are applied in the relative nodes and the loads are applied on the various types of elementary structures that make up the analysed ship structure previously defined.
Using code 13 the structure, defined at an earlier stage, can be displayed on the screen to see any input errors and correct them, with code 14 the static finite element analysis is started and a text document is created with the corresponding results and using code 15 the software plots the results of the analysis and the data entered by the user. Finally, with code 16 is possible to have a graphical display of the results and create a text file from which the structural checks can be done with the ones of other softwares.
Throughout the procedure, the software generates a series of files of various formats, from the title of the initial project and in the folder previously defined, necessary for the DOS solver to enter, modify, view and perform the calculations required in the various steps of the project and the various results output in tabular and printable form.

These files are readable with a text file program such as notepad.

4.3 GEOMETRIC DESCRIPTION AND DISCRETIZATION

The project data provided by the university were the main features of the ship analysed and the construction plan of the new bow. The data provided show that the bow has a transverse structure and has diaphragms interspersed with frames, the structure is heavily reinforced to withstand the huge forward loads to which the ship is subjected. In fact, the frame space is very small, and the ordinary elements are transverse as the local loads are bigger than the global loads in this area. With the tables provided by the university, it was possible to define a digital three-dimensional model of the bulbous bow through the geometric modelling software Rhinoceros [10] as shown in the figure.

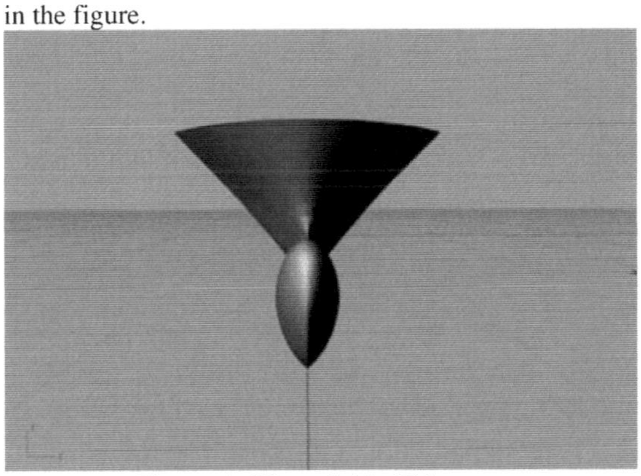

Figure 4: 3D model in Rhinoceros

When performing a FEM analysis, it is important to evaluate the boundary conditions, and therefore the orientation of the structure is also important when performing the analysis. In this case the structure has been positioned in such a way as to have the x axis in the longitudinal stern-bow direction, the y axis is in the direction of the breadth and the vertical z axis is positive upwards (see Figure 4). Subsequently, the 3D model was intersected with various planes parallel to the y-z plane with a regular interval of 600mm and, from the intersection of the figure with the aforementioned planes, some points of interest were obtained. The other notable points were obtained by further intersecting the model with horizontal planes (water lines).

```
PASEF ©  NUMERO DI NODI     = 1478
         NUMERO DI ASTE     = 0
         NUMERO DI TRAVI    = 0
         NUMERO DI LASTRE   = 0
         NUMERO DI PIASTRE  = 1549
         NUMERO DI SOLIDI   = 0
         ACC. GRAVITA' X    = 0
         ACC. GRAVITA' Y    = 0
         ACC. GRAVITA' Z    =-9.81
```

Figure 5: PASEF – Input data

Therefore, for the purpose of the geometric discretization of the real 3D model of the forward structure, 1478 nodes (strategic points of the structure) were obtained. These nodes were imported into the PASEF platform using the respective x, y and z coordinates, and the degrees of freedom of each node. Subsequently, using the plate element, the initial mesh was reproduced in the form of a polysurface made up of 1549 plates. The high number of nodes and plates used, allows to perform a detailed FEM analysis closer to the real model.

4.4 CONTRAINTS AND MATERIALS

The structure analysed is only the forward part of a ship (complex structure). It was therefore decided to constrain it through the nodes of the collision bulkhead, hypothesized to be welded to the cylindrical body to restore continuity with the rest of the ship. Then the various nodes that define the structure, the relative degrees of freedom and the orientation of the structure with respect to the Cartesian reference system are inserted through the definition of the gravity acceleration vector with its three components. When the input data are set into PASEF, an .INO file is generated which contains the above data.

```
BOW - Blocco note
File  Modifica  Formato  Visualizza  ?
1478    0     0     0     1549    0    0    0     -9.81
0.000     5.733    10.000    123456
0.600     5.562    10.000    0
1.200     5.389    10.000    0
1.800     5.213    10.000    0
2.400     5.030    9.998    0
3.000     4.840    9.998    0
3.600     4.644    9.998    0
4.200     4.442    9.998    0
4.800     4.229    9.998    0
5.400     4.002    9.998    0
6.000     3.757    9.998    0
6.600     3.491    9.999    0
7.200     3.206    9.999    0
7.800     2.908    9.998    0
8.400     2.579    9.998    0
9.000     2.178    9.998    0
9.600     1.617    9.999    0
10.200     0.620     9.999    0
10.315     0.000     9.999    0
10.200     -0.620     9.999    0
9.600     -1.617     9.999    0
9.000     -2.178     9.998    0
```

Figure 6: PASEF – nodes and constrains

Subsequently, the program requests the definition of the material of which the elements of the structure are made with their properties. The material chosen for the various structures is high strength steel HSS whose properties are $E = 2.10x10^{11} N/m^2$, $\nu = 0.3$, $k = 0$, $\rho = 7.85x10^3 Kg/m^3$. The plates of this material (HSS) were used both for the construction of the planking panels and for the reinforcements such as collision bulkhead, web, bridges and longitudinal bulkhead (at the centre line). Therefore, the plates of the structure have been divided into groups based on the intended use and thickness, as they will have different thicknesses depending on the intended use. For the planking a thickness of 13mm (group 1) was adopted, for the collision bulkhead a thickness of 15mm (group 2), for the longitudinal bulkhead a thickness of 15mm (group 9), for the diaphragms a thickness of 11mm (group 3), for the last watertight bridge a thickness of 15mm was adopted, for the 4th bridge a thickness of 7.5mm and for the remaining bridges a thickness of 11mm (groups 6, 7, 8).

Once the plates have been inserted and with their material and their properties and thickness, PASEF generates two files:

- .IPP plate properties input.
- .IPI input plates with coordinates of the 4 nodes at the vertices of the element.

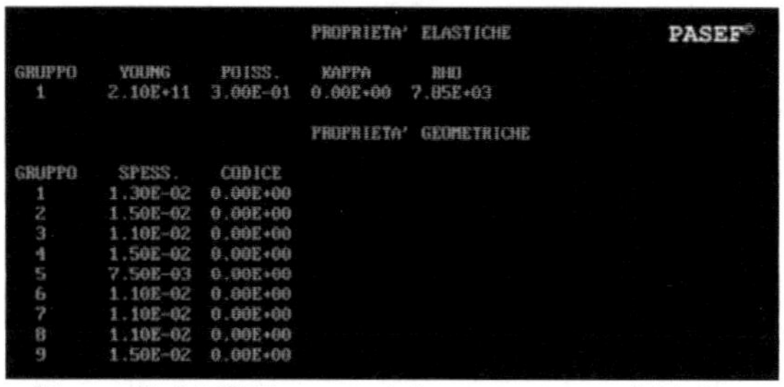

PASEF©

PROPRIETA' ELASTICHE

GRUPPO	YOUNG	POISS.	KAPPA	RHO
1	2.10E+11	3.00E-01	0.00E+00	7.85E+03

PROPRIETA' GEOMETRICHE

GRUPPO	SPESS.	CODICE
1	1.30E-02	0.00E+00
2	1.50E-02	0.00E+00
3	1.10E-02	0.00E+00
4	1.50E-02	0.00E+00
5	7.50E-03	0.00E+00
6	1.10E-02	0.00E+00
7	1.10E-02	0.00E+00
8	1.10E-02	0.00E+00
9	1.50E-02	0.00E+00

Figure 7: PASEF – material: properties and thickness

Once the input data has been entered in PASEF, the program outputs the following graphical representations.

4.5 STRUCTURAL STRESSES

Defined the geometry, the material of the structure and its properties, the next step in the study was the evaluation of the stresses on the structure, i.e. the nodal forces. For the choice of loads to be applied to the structure, being a static analysis, the weight and hydrostatic pressure applied to the nodes located below the design waterline at 5.7 m from the Base Line were considered [10].

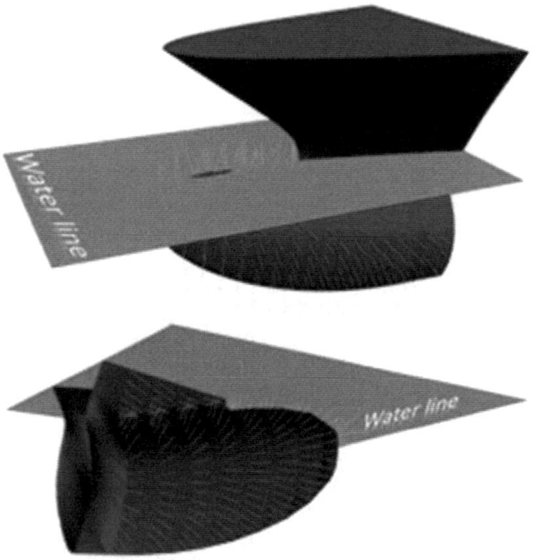

Figure 8: 3D modelling of the bow with hydrostatic pressure

The hydrostatic pressure at the nodes is given by the relation:

$P_S = \rho g (T - z)$ [N/m²]

It must be multiplied by the area of each node, to obtain the pressure force F from the formula:

$F = P_S * A$ [N]

in module and its components F_x, F_y, F_z through the director cosines.

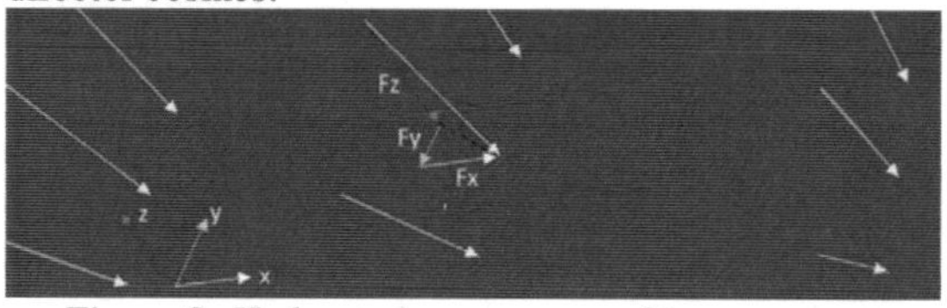

Figure 9: Hydrostatic pressure and its components

Since the discretization is equally spaced, as described in the previous paragraphs, it has been calculated that on average the area on which the pressure p acts for each node is equal to 0.313 m².

The program calculates the weight of the structure automatically from the data relating to the geometry of the structure and the properties of the material previously set. The forces F were applied on the normal to the plate in each node and the module and the relative components F_x, F_y, F_z were determined using the Excel program. These components are automatically inserted by the program in a .IFN file (input nodal forces).

4.6 STATIC ANALYSIS

Once the Static Analysis has started, the PASEF program generates a text file containing the number of equations to be solved and unknowns divided into blocks of matrices. It also generates temporary files in which it solves these equations.

At the end of the numerical calculation the results are provided by the program in 3 files:

- .REA containing the components of the support reactions on all the constrained nodes;
- .SPO containing the nodal displacements caused by the applied forces;
- .STR containing the new coordinates of the nodes and the structure, after deformation.

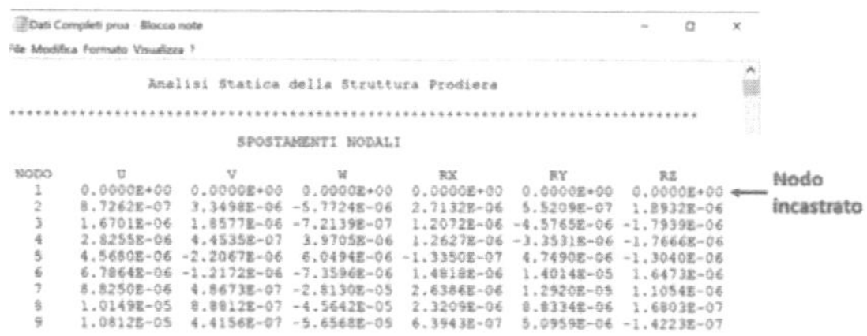

Dati Completi prua - Blocco note

File Modifica Formato Visualizza ?

Analisi Statica della Struttura Prodiera

SPOSTAMENTI NODALI

NODO	U	V	W	RX	RY	RZ
1	0.0000E+00	0.0000E+00	0.0000E+00	0.0000E+00	0.0000E+00	0.0000E+00
2	8.7262E-07	3.3498E-06	-5.7724E-06	2.7132E-06	5.5209E-07	1.8932E-06
3	1.6701E-06	1.8577E-06	-7.2139E-07	1.2072E-06	-4.5765E-06	-1.7939E-06
4	2.8255E-06	4.4535E-07	3.9705E-06	1.2627E-06	-3.3531E-06	-1.7666E-06
5	4.5680E-06	-2.2067E-06	6.0494E-06	-1.3350E-07	4.7490E-06	-1.3040E-06
6	6.7864E-06	-1.2172E-06	-7.3596E-06	1.4818E-06	1.4014E-05	1.6473E-06
7	8.8250E-06	4.8673E-07	-2.8130E-05	2.6386E-06	1.2920E-05	1.1054E-06
8	1.0149E-05	8.8812E-07	-4.5642E-05	2.3209E-06	8.8334E-06	1.6803E-07
9	1.0812E-05	4.4156E-07	-5.6568E-05	6.3943E-07	5.0959E-06	-1.4223E-07

Nodo incastrato

Figure 10: PASEF – static analysis, results

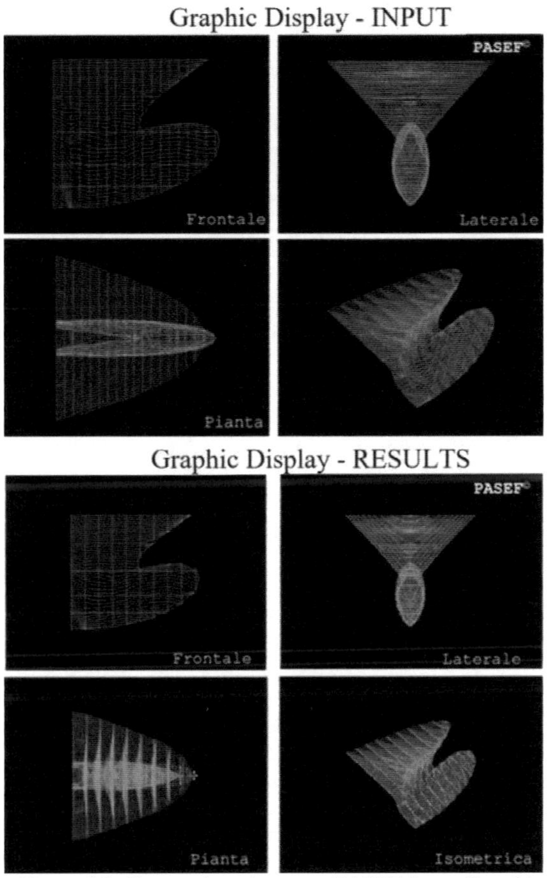

Figure 11: Comparison between PASEF Input Graphic Display and Graphic Display of Results

4.7 STRUCTURAL INCREASES RELATED TO THE SHIP'S MISSION PROFILE

From RINA, P.te F, Chap. 9, Class of ice 'ICE', is possible to define the thickness of the planking for navigation between the ice, that must be not less than the value obtained from the following formula, for transverse structures:

$T = 667s\ \mathrm{radq}(F_1 p_{pl}/R_{eH}) + t_c$

Where:

$F_1 = 1{,}3 - 4{,}2/((h/s) + 1{,}8)^2$

$p_{pl} = 0{,}75p$.

4.8 NODAL SHIFT DIAGRAMS

Following the execution of PASEF, the program returns the results of the nodal displacements obtained as a result of the deformation. These results are shown here in graphical form using Excel for each of the 770 nodes:

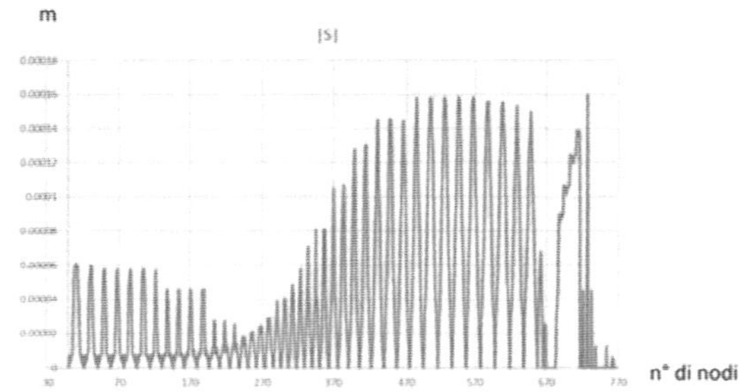

Component of nodal displacements along x:

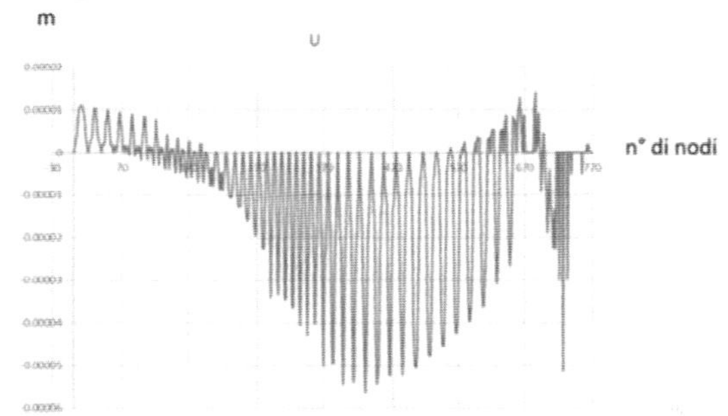

Component of nodal displacements along y:

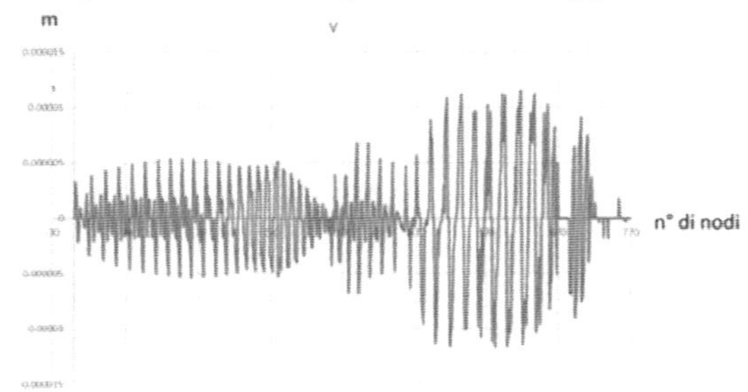

Component of nodal displacements along z:

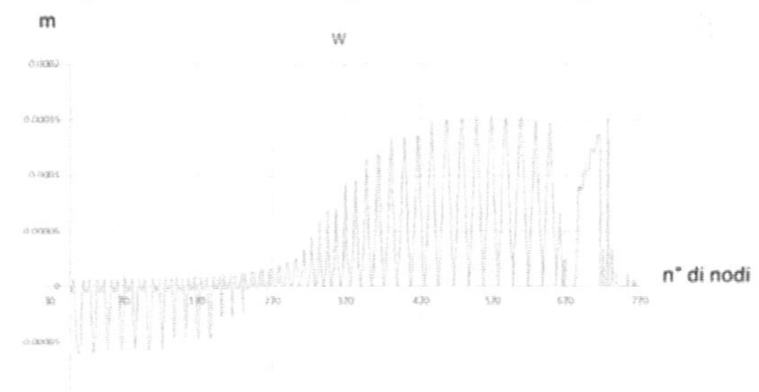

Figure 12: results of the static analysis

The maximum shifts result in a minimum value, near to 0. The figures have been modified in the graphic representation to emphasize the parts of the ship that are most subject to stresses.

4.9 THE SAFETY FACTOR

Nowadays, safety plays an important role in many areas, especially the engineering one.

Therefore, a role of fundamental importance in any engineering calculation, both for design and for check, is represented by the so-called «safety factor».

Regarding the plasticization limit in a point, the Safety factor for a certain stress state is given by the amplification factor (positive) to be applied to the stress state so that the plasticization limit is reached. The safety factor can be determined as follow:

$\sigma_{eq}(\gamma\sigma_1, \gamma\sigma_2, \gamma\sigma_3) = \gamma\sigma_{eq}(\sigma_1, \sigma_2, \sigma_3) = \sigma_0 \rightarrow \gamma = \sigma_0/\sigma_{eq}$

It results that the safety factor γ must be minor or equal to the obtained stress.

$\sigma_{eq} \leq \sigma_{amm} \rightarrow \sigma_0/\sigma \leq \sigma_0/\gamma \rightarrow \sigma \geq \gamma$

The written report is the basis of the so-called "Method of Check of Admissible Stresses". The required safety factor is always greater than the unit (> 1) and depends on many factors depending on the material used, the loads, and the approximations of the mathematical modelling. For many years, the security approach of engineering structures has relied on deterministic security factors, also based on experience. This has often caused optimization problems or incorrect assessments of factors, potentially capable of threatening the structural integrity. In recent years, statistical techniques that have been introduced are able to consider the intrinsic variability of some important factors (resistance of materials, etc.) In general, the safety coefficient can also be expressed in this way:

$\gamma = \beta^*\eta / C_{st}$

4.10 STRUCTURAL CHECK OF THE BOW

To check if a structure is in safe conditions, the value of the equivalent stress is calculated and compared with the admissible stress by:

$\sigma_{eq} \leq \sigma_{amm}$

Using the Hencky - Von Mises criterion for which:

$\sigma_{eq} = [\sigma_1^2 + \sigma_2^2 + \sigma_3^2 - (\sigma_1\sigma_2 + \sigma_2\sigma_3 + \sigma_3\sigma_1) + 3(\tau^2_{xy} + \tau^2_{yz} + \tau^2_{zx})^{1/2}$ [MPa]

and knowing that

$\sigma_{amm} = 175/k$ [MPa]

where k is a non dimensional coefficient, function of the minimum yield stress R_{eH} of the material, thus defined by RINA, P.te B, Chap.4, Section 1:

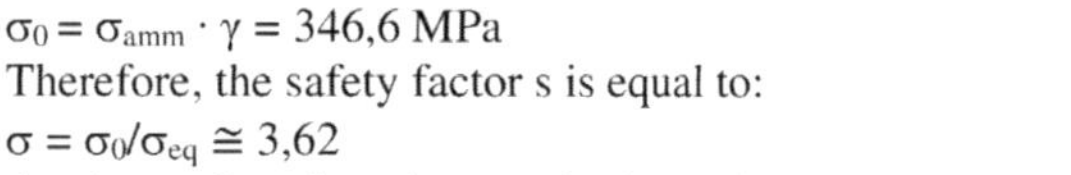

Figure 13: Safety factor, table and results

5. CONCLUSIONS

Calculation of the Safety factor is

Given γ = 1.5 (Safety factor γ required by the project in this case) it results that:

$\sigma_0 = \sigma_{amm} \cdot \gamma = 346{,}6$ MPa

Therefore, the safety factor s is equal to:

$\sigma = \sigma_0/\sigma_{eq} \cong 3{,}62$

that is, not less than that required γ = 1.5.

In case of a complete structural analysis, considering all the forces involved, the level of safety will be different. Having considered only the static loads, which are only a small part of the external loads acting on the structure, an acceptable level of safety has been obtained.

The results obtained from the FEA have been compared with the ones obtained with the RINA rules.

The thicknesses obtained with the rules result to be bigger than the ones obtained with the FEA, not only because the loads considered are the static ones, but also because the rules adopt a more conservative approach, based also on an empirical advance.

6. ACKNOWLEDGEMENTS

The authors are grateful for the software and helpful advices from Prof. Francesco Russo Spena to write this paper.

7. REFERENCES

1. RINA 2017
2. S. KHASNABIS, 'Types of Bow Designs Used For Ships', *www.marineinsight.com*, 2020.
3. S. CHACKRABORTY, 'What's the importance of bulbous bow?', *www.marineinsight.com*, 2020.
4. S. KHASNABIS, 'Nose-Job for Ships', *www.marineinsight.com*, 2020.
5. G. RUSSO KRAUSS, 'Appunti di Progetto della Nave', *Università degli Studi di Napoli "Federico II"*, 1990.
6. K. RISKA, 'Design of Ice Breaking Vessels', 2010.
7. RINA – Additional Class Notations
8. F. RUSSO SPENA, 'Appunti di Scienza delle Costruzioni II, *Università degli Studi di Napoli "Federico II"*, 2000.
9. PASEF, *FEM Software*
10. RHINOCEROS, *3D modelling software.*

8. AUTHORS BIOGRAPHY

Benedetta D'Amato Naval Architect Msc, associated member of the Royal Institution of Naval Architects, has reached the qualification of Professional Engineer.
During the studies she has been involved in a FEM team project and carried out research studies for hybrid propulsion systems with fuel cells.

Giuseppe Soriano Naval Architecture Student, after the BSc, he is actually completing the Master Degree at the Università di Napoli "Federico II". He has been an intern at the SARC BV (NL) in 2019 for the Erasmus programme. He was part of the team for the FEM analysis project.

Andrea Ambrosio Naval Architect MSc, actually employed at Fincantieri covering the role of Assistant Head outfitting. He has gained experience at the Palumbo Shipyard in refitting operations. His master thesis was a project on fuel cells installed on a cruise ship and, during the studies, he has been involved in the team project for FEM analysis.

INVESTIGATION OF STRUCTURAL STRENGTH OF ICE REINFORCED SIDE SHELL STRUCTURE UNDER FINNISH SWEDISH ICE CLASS NOTATIONS

S. Rahman, Structural Engineer, Ankerbeer Engineering and Design, The Netherlands.

SUMMARY

The melting ice in Arctic Ocean creates more shipping possibilities in Northern Sea Routes (NSR) which offers less fuel consumption and less voyage time of passing through the Arctic Ocean than the south one. This paper investigates the structural strength of various configurations of ice reinforced side shell structure under Finnish Swedish ice class notations. In order to compare among different ice belt configurations, a programme is developed. This programme uses vessel configuration with Finnish Swedish ice class rule criteria to calculate the strength requirements of ice belt structure. Different material grade, stiffener spacing, frame spacing and type of framing system are used in order to investigate the influence on lightship weight and structural requirements of ice belt structure. Finally, a set of recommendations are proposed which offers less lightship weight and structural requirements of ice belt structure.

NOMENCLATURE

FSICR	Finnish –Swedish Ice Class Rules
IACS	The International Association of Classification Societies
CSR	Common structural rule
UIWL	Upper ice waterline
LIWL	Lower ice waterline
Zp	Plastic section modulus
Ze	Elastic section modulus
S	Frame spacing (m)
P	Design pressure (MPa)
σ_y	Material yield stress (MPa)
h	Height of load area (m)
l	Span of the frame (m)

1. INTRODUCTION

The demand for shipping in Arctic areas is increasing due to it's shorter voyage time. Due to global warming, ice in Arctic areas is melting and it creates more opportunities for shipping and exploring natural resources in Arctic areas. Generally, polar materials are used for the construction of ice class vessels which is different than normal ship building steel and has the low temperature effect on structural strength. In ice class vessels, ice belt on side shell is built with heavier scantling which reduces the impact load from ice.

Now a days there are several regulations for designing the ice belt structure. These regulations are proposed from numerical analysis and practical experience during navigation in ice to ensure the structural strength [1]. Most classification societies have adopted Finnish – Swedish Ice Class Rules (FSICR) for designing the ice belt structure [2]. FSICR will be applicable only to first year ice conditions in the Northern Baltic region [3]. The ice in the Northern Baltic region is stronger due to its low salt content. In 2006, The International Association of Classification Societies (IACS) developed unified requirements (IACS UR) for ice class vessels. The difference between different classification society's ice belt strength requirements are based on different theories, constraints and approaches [4]. FSICR is based on elastic method while IACS UR is based on plastic method. Daley et.al. [5] had mentioned that a plastic method provides better material distribution to resist extreme ice loads. It also provides safety margins to hull girder collapse. Through plastic method allows some local deformations but it provides enough strength to the ice belt structure which is not possible in elastic method. In order to compare between elastic and plastic section modulus, Riska [2] proposed a shape factor (Zp/Ze=1.35) which gives a rough comparison. This shape factor is based on bulb profiles with attached plating thickness of 15mm and width of 600mm for elastic and plastic method respectively. The main goal of this paper is to compute the local strength of ice belt structure under different structural configuration and material grade. In this analysis, FSICR is used to verify the local strength of ice belt structure. Hence, a parametric study is performed by changing main parameters like stiffener spacing, frame spacing and material grade in order to get the influence of weight and structural properties within the ice belt structure. For that a programme is developed in Microsoft excel which validates the ice belt structural strength calculation with FSICR.

2. ICE CLASS NOTATIONS

Ice class notations depend on the navigational area of the vessel. Then it's up to owner's decision to choose an appropriate ice class notation for the vessel. FSICR is based on the elastic approach and it uses engine power and vessels displacement for scantling of ice belt structure.

Table 1: Ice class notation (FSICR) [6]

Ice notation	Detail	Ice thickness (m)
I A super	Ships with such structure, engine output are generally capable of navigating in difficult ice conditions without the assistance of ice breakers.	1m
IA	Ships with such structure,	0.8m

	engine output are capable of navigating in difficult ice conditions with the assistance of ice breakers when necessary.	
IB	Ships with such structure, engine output are capable of navigating in moderate ice conditions with the assistance of ice breakers when necessary.	0.6m
IC	Ships with such structure, engine output are capable of navigating in light ice conditions with the assistance of ice breakers when necessary.	0.4m

Hull area extension of ice belt region is defined by upper ice waterline (UIWL) and lower ice waterline (LIWL). So, the ice class vessels draft including trim will be within UIWL and LIWL for all loading conditions. In FSICR, ice belt regions are divided in to three areas; the fore body, the mid body and the aft body. In this analysis, structural strength of all three areas are investigated. Flat side area is important for defining ice belt region. Mid body area in ice belt is defined from the aft boundary of the bow area to a line parallel to and 0.04L aft of the flat side line where the waterlines run parallel to the centreline [6].

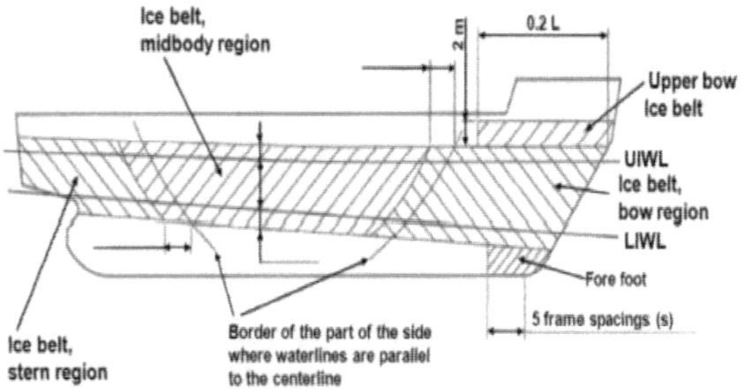

Figure 1: Ice belt area (FSICR) [6]

Hull extension at ice belt area is required for shell plating and frames. The minimum vertical extension of shell plating is shown in Table 2. Also the minimum vertical extension of ice frame is shown in Table 3.

Table 2: Vertical extension of shell plating (FSICR) [6]

Ice class	Hull region	Above UIWL	Below LIWL
IA Super	Bow	0.60 m	1.20 m
	Midbody		
	Stern		1.0 m
IA	Bow	0.50 m	0.90 m
	Midbody		0.75 m
	Stern		
IB and IC	Bow	0.40 m	0.70 m
	Midbody		0.60 m
	Stern		

Table 3: Vertical extension of ice framing (FSICR) [6]

Ice class	Hull region	Above UIWL	Below LIWL
IA Super	Bow	1.2 m	Down to tank top or below top of the floors
	Midbody		2.0 m
	Stern		1.6 m
IA, IB and IC	Bow	1.0 m	1.6 m
	Midbody		1.3 m
	Stern		1.0 m

3. LOCAL STRENGTH

In order to calculate the local strength of ice belt structure, firstly, it is necessary to calculate the design pressure due to ice. In FSICR, the design pressure due to ice depends upon the ice thickness. After determining the ice load then ice belt plating thickness and the shear area and section modulus of ice framing is calculated. In FSICR, corrosion allowance (t_c) is taken as 2mm.

3.1 Plate thickness in ice belt

For transverse framing, the thickness of the shell plating is determined by the formula [6]:

$$t = 667s\sqrt{\frac{f_1 p_{pl}}{\sigma_y}} + t_c \ [\mathrm{mm}].$$

and for longitudinal framing, the thickness of the shell plating is determined by the formula [6]:

$$t = 667s\sqrt{\frac{p}{f_2 \sigma_y}} + t_c \ [\mathrm{mm}]$$

Where, P_{pl}= 0.75*P (MPa)

$$f_1 = 1.3 - \frac{4.2}{(h/s+1.8)^2}, \text{ maximum } 1.0,$$

$$f_2 = \begin{cases} 0.6 + \frac{0.4}{h/s}, when\ h/s \leq 1 \\ 1.4 - 0.4(h/s), when\ 1 \leq h/s \leq 1.8, \end{cases}$$

and h is found from Table 4.

Table 4:h_i and h value for different ice class noation [6]

Ice class notation	h_i (m)	h (m)
IA Super	1	0.35
IA	0.8	0.3
IB	0.6	0.25
IC	0.4	0.22

3.2 Section modulus and shear area of frames in ice belt

The section modulus of transverse frame is calculated using the formula [6]:

$$Z = \frac{pshl}{m_t \sigma_y} 10^6 \ [\mathrm{cm}^3]$$

and the effective shear area for transverse frame is calculated from the following formula [6]:

$$A = \frac{\sqrt{3} f_3 p h s}{2\sigma_y} 10^4 \, [\mathrm{cm}^2]$$

Where,

f_3=1.2

$$m_t = \frac{7m_o}{7-5h/l'}$$

m_o= 6 (Frames extending from the tank top to the main deck of a single-decked vessel)

The section modulus of a longitudinal frame is calculated using the formula [6]:

$$Z = \frac{f_4 p h l^2}{m \sigma_y} 10^6 \, [\mathrm{cm}^3]$$

The effective shear area of a longitudinal frame is [7]:

$$A = \frac{\sqrt{3} f_4 f_5 p h l}{2\sigma_y} 10^4 [\mathrm{cm}^2]$$

Where,

f_5=2.16

$f_4 = (1 - 0.2\, h/s)$

4. SENSITIVITY ANALYSIS

In order to investigate the optimized ice belt structure a sensitivity analysis is performed. To perform the sensitivity analysis, a programme is developed in Microsoft excel which calculates the scantling of ice belt structure using Finnish Swedish ice class rule. Different material grade, stiffener spacing, frame spacing and type of framing system are used in order to investigate the influence on lightship weight and structural requirements of ice belt structure. While calculating the ice belt structure it is kept in mind that common structural rules are also fulfilled.

4.1 Effect of stiffener spacing on ice belt structure

In this analysis, in order to investigate the effect of stiffener spacing longitudinal framing system is adopted. In this analyis, web frame spacing is kept constant which is 2.1m while the stiffener spacing is changing from 300mm to 500mm.

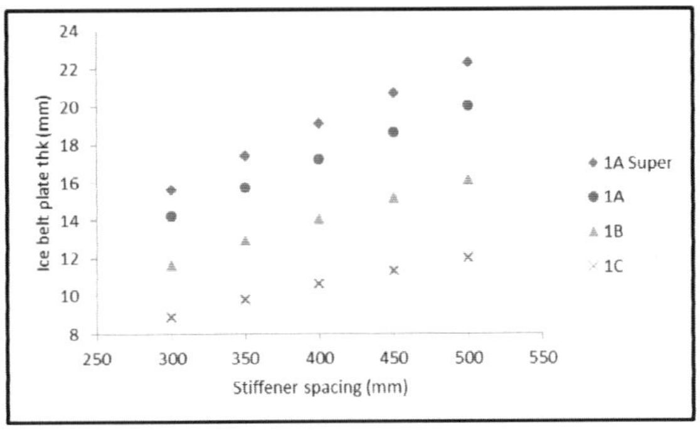

Figure 2: Ice plt thk vs stiffener spacing (Aft body)

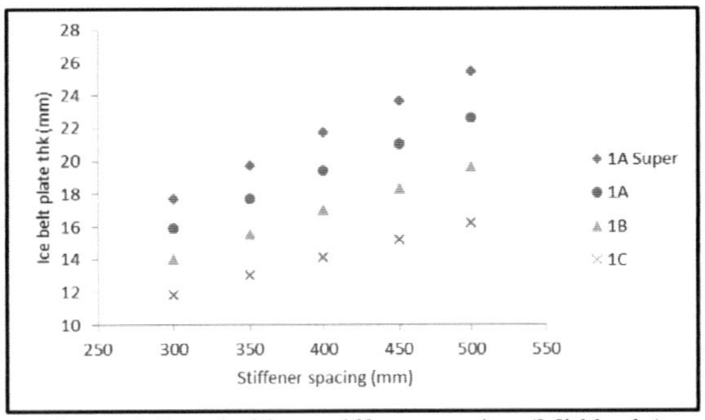

Figure 3: Ice plt thk vs stiffener spacing (Mid body)

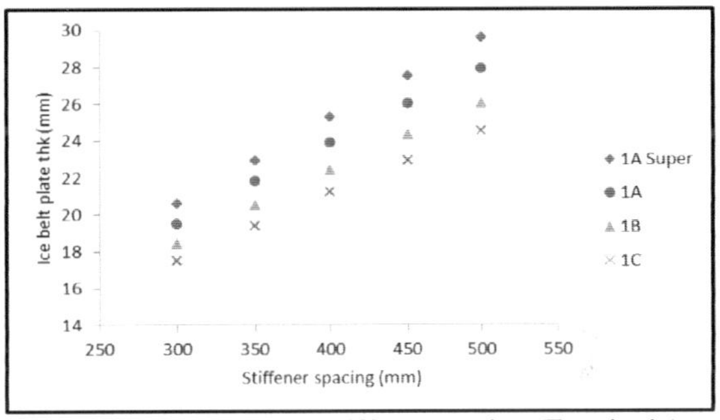

Figure 4: Ice plt thk vs stiffener spacing (Fore body)

From Figures 2, 3 and 4 it is observed that ice plating thickness increases with the increase of stiffener spacing and the increase is almost linear. Moreover, it is found that ice belt plating thickness increases almost 10% with increasing stiffener spacing by 50mm. Also, it is noticed that ice class 1A super needs higher scantling than 1C due to high ice load.

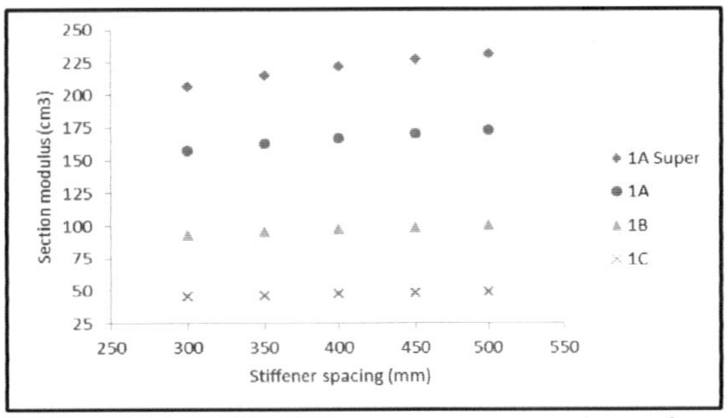

Figure 5: Section modulus vs stiffener spacing (Aft body)

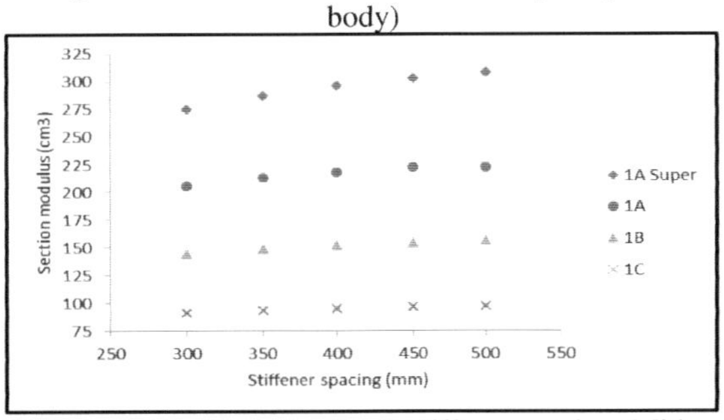

Figure 6: Section modulus vs stiffener spacing (Mid body)

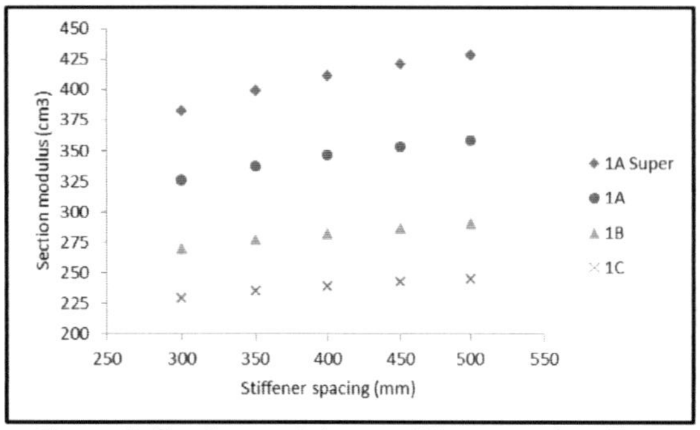

Figure 7: Section modulus vs stiffener spacing (Fore body)

From Figures 5, 6 and 7, it is seen that ice framing section modulus does not increase so much with the increase of stiffener spacing and the increase is almost 5% with increasing stiffener spacing by 50mm.

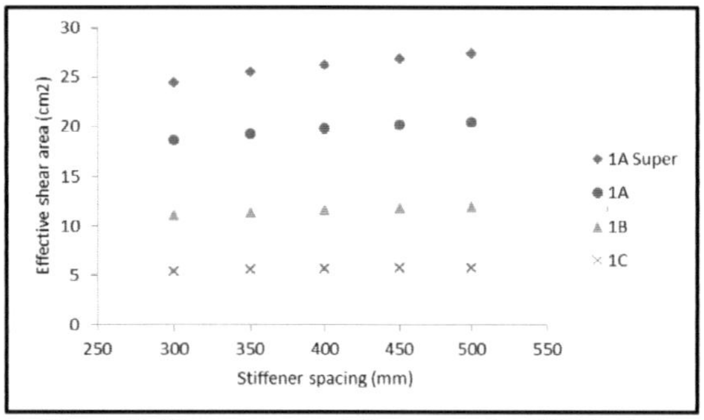

Figure 8: Shear area vs stiffener spacing (Aft body)

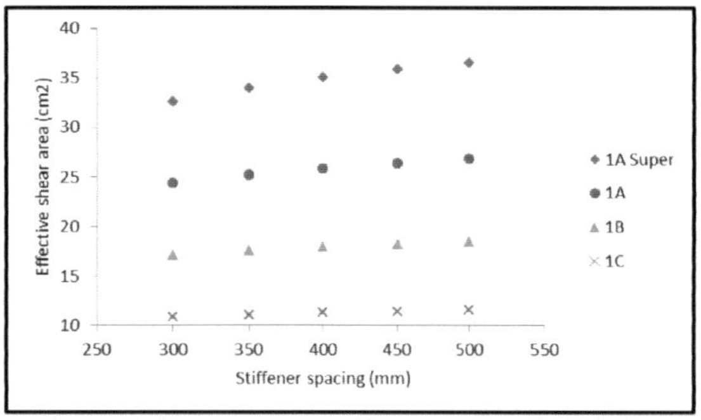

Figure 9: Shear area vs stiffener spacing (Mid body)

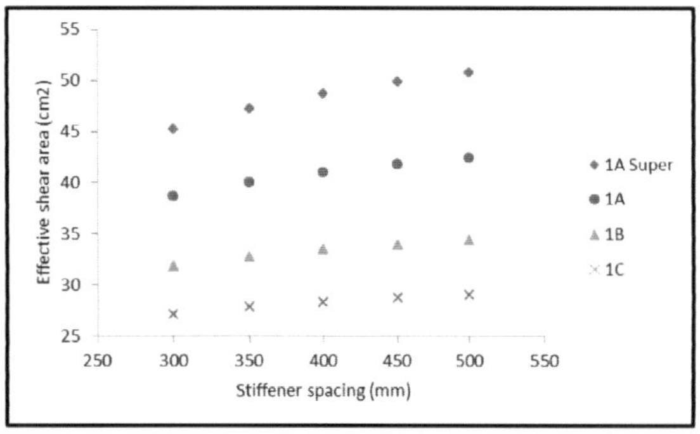

Figure 10: Shear area vs stiffener spacing (Fore body)

From Figures 8, 9 and 10, it is seen that ice framing effective shear area does not increase so much with the increase of stiffener spacing and the increase is almost 5% with increasing stiffener spacing by 50mm.

4.2 Effect of frame spacing on ice belt structure

In this analysis, in order to investigate the effect of frame spacing longitudinal framing system is adopted. In this analyis, stiffener spacing is kept constant which is 350mm while the frame spacing is changing from 1.2m to 2.4m.

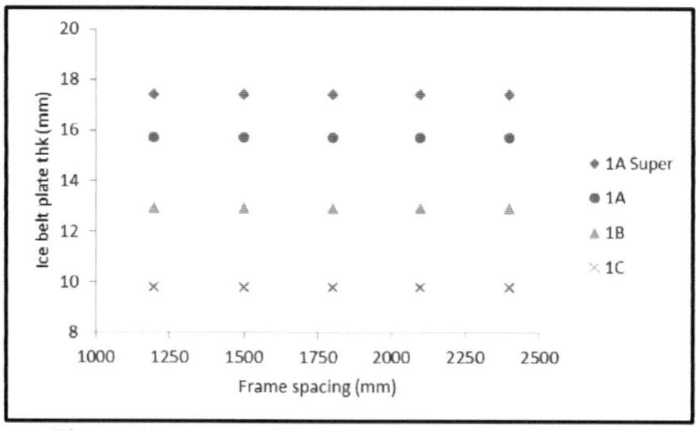

Figure 11: Ice plt thk vs frame spacing (Aft body)

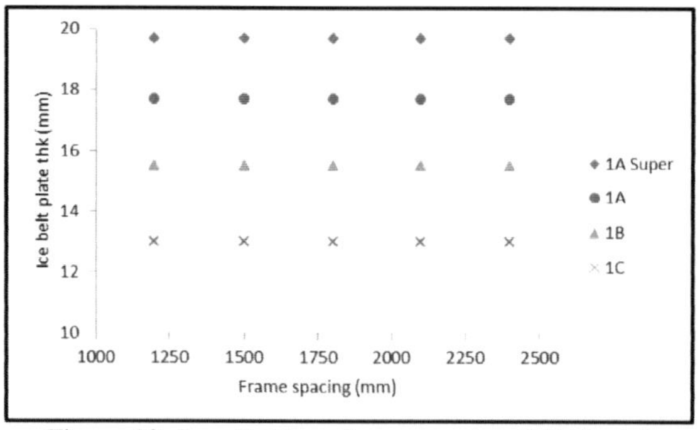

Figure 12: Ice plt thk vs frame spacing (Mid body)

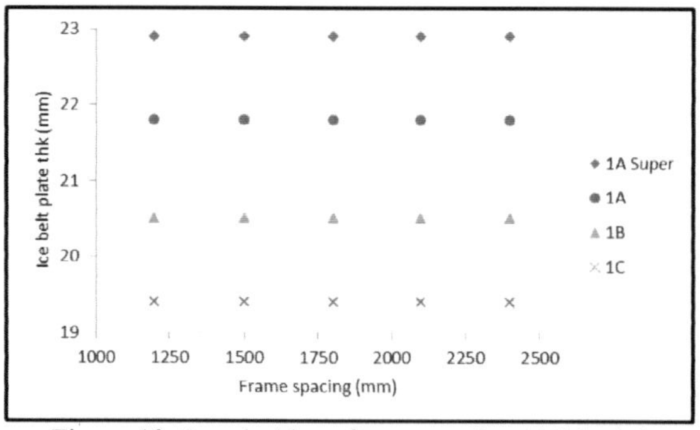

Figure 13: Ice plt thk vs frame spacing (Aft body)

From Figures 11, 12 and 13, it is investigated that ice belt plating thickness does not change with the increase of frame spacing.

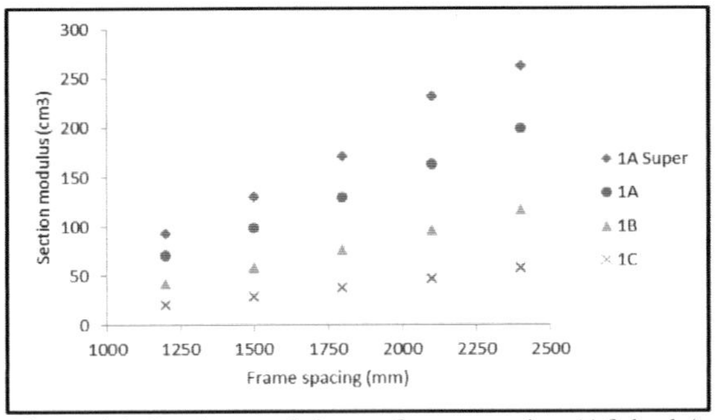

Figure 14: Section modulus vs frame spacing (Aft body)

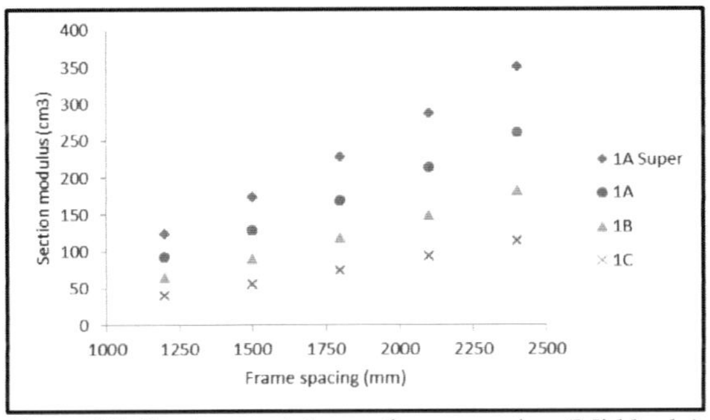

Figure 15: Section modulus vs frame spacing (Mid body)

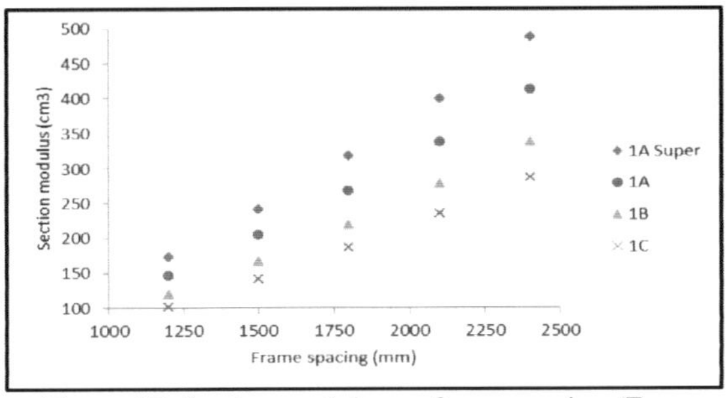

Figure 16: Section modulus vs frame spacing (Fore body)

From Figures 14, 15 and 16, it is observed that ice framing section modulus increases with the increase of frame spacing and the increase is almost linear. Moreover, it is found that ice framing section modulus increases almost 25% with increasing frame spacing by 300mm.

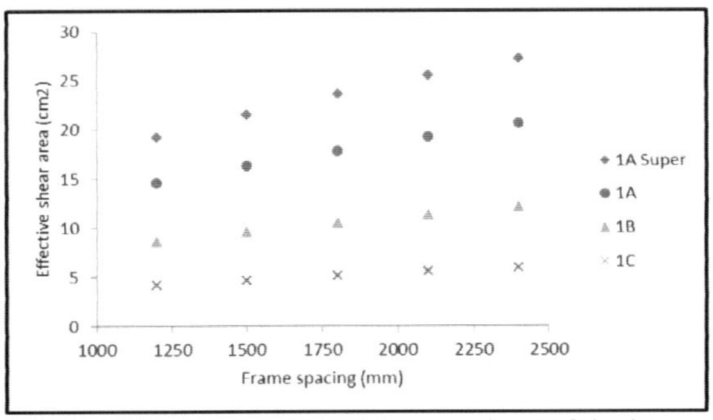

Figure 17: Shear area vs frame spacing (Aft body)

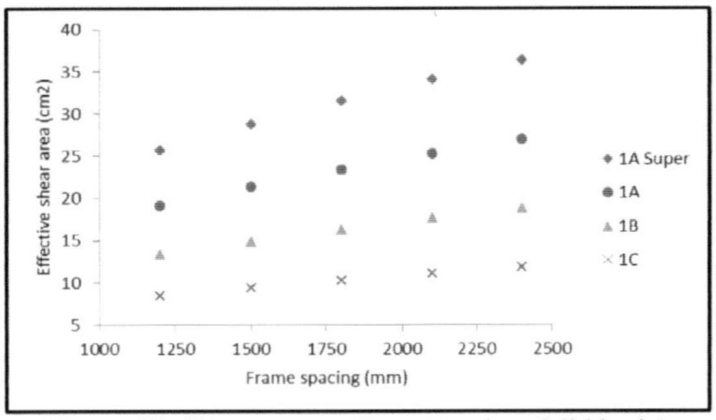

Figure 18: Shear area vs frame spacing (Mid body)

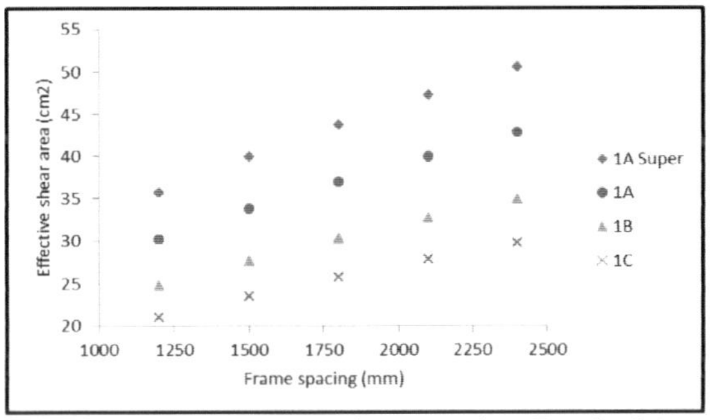

Figure 19: Shear area vs frame spacing (Fore body)

From Figures 17, 18 and 19, it is seen that ice framing effective shear area does not increase so much with the increase of frame spacing. Moreover, it is seen that ice framing shear area increases almost 10% with increasing frame spacing by 300mm.

4.3 Effect of type of framing system on ice belt structure

Framing system is an important parameter for defining the scantling of ice belt structure. In this analysis, transverse and longitudinal framing system is considered. Also, stiffener spacing and web frame spacing is kept constant which is 350mm and 2.1m respectively.

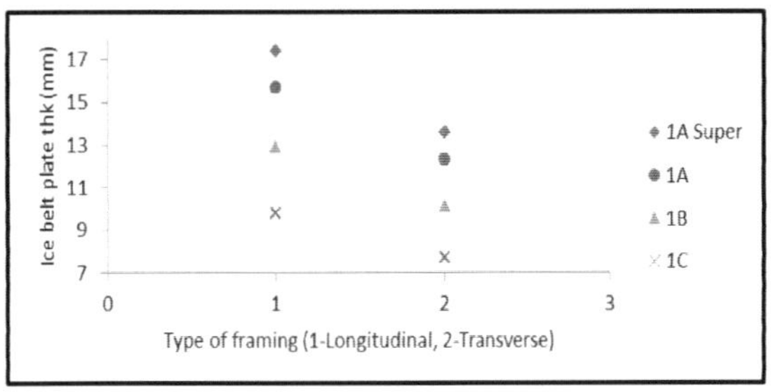

Figure 20: Ice plt thk vs framing system (Aft body)

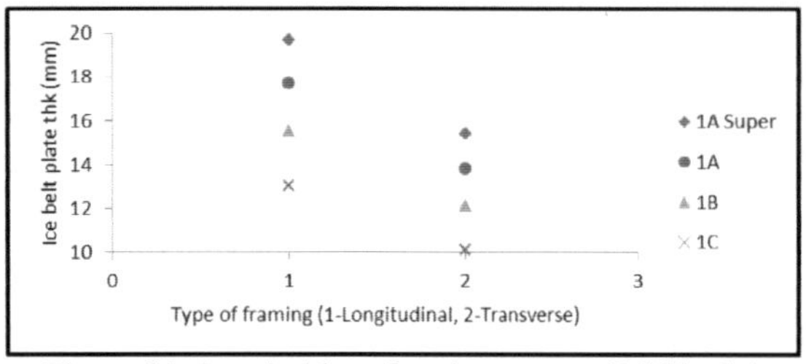

Figure 21: Ice plt thk vs framing system (Mid body)

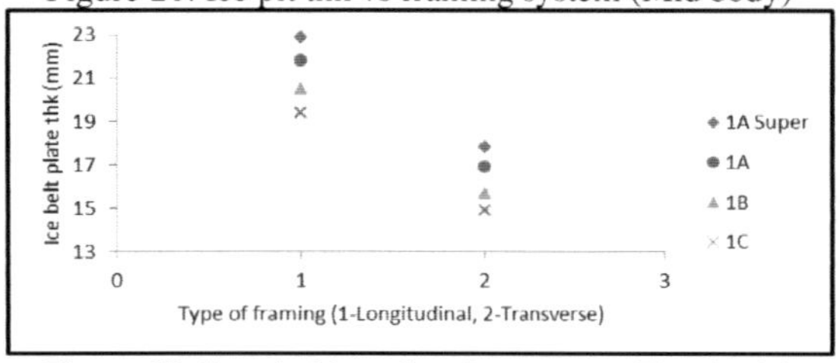

Figure 22: Ice plt thk vs framing system (Fore body)

From Figures 20, 21 and 22, it is found that longitudinal framing system gives higher ice plating thickness than transverse framing system which is almost 20%.

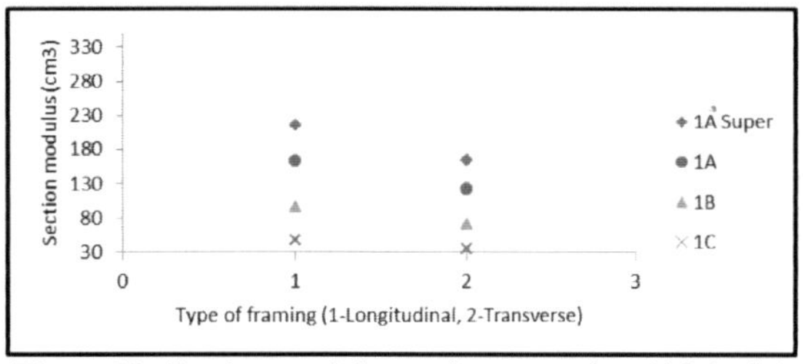

Figure 23: Section modulus vs frame spacing (Aft body)

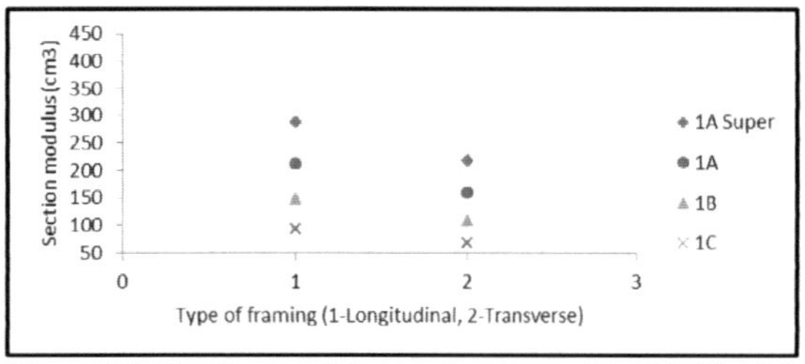

Figure 24: Section modulus vs frame spacing (Mid body)

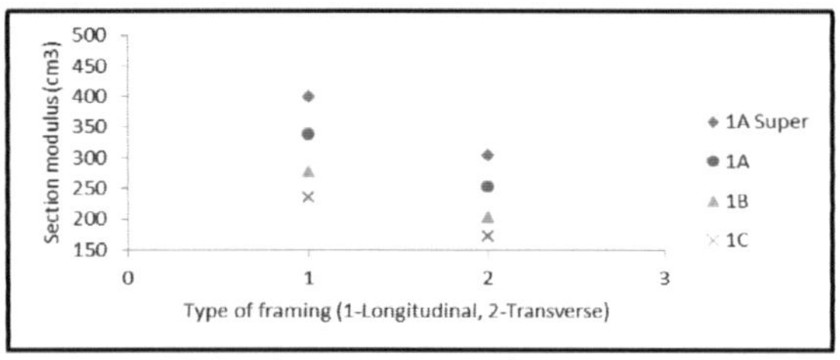

Figure 25: Section modulus vs frame spacing (Fore body)

From Figures 23, 24 and 25, it is seen that longitudinal framing system gives higher ice framing section modulus than transverse framing system which is almost 25%.

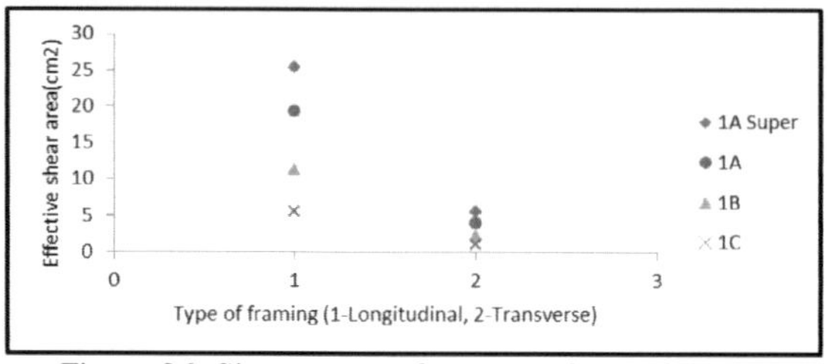

Figure 26: Shear area vs frame spacing (Aft body)

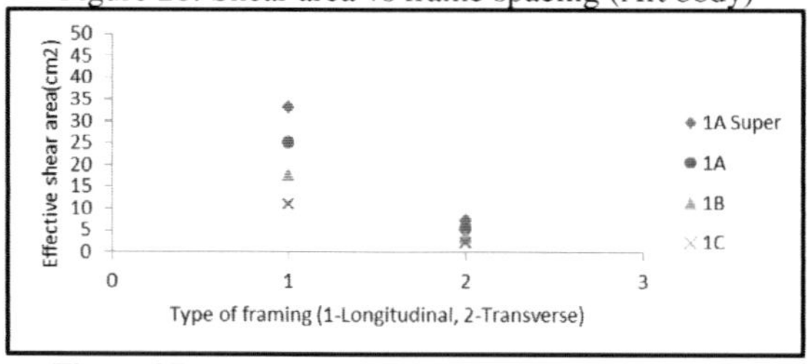

Figure 27: Shear area vs frame spacing (Mid body)

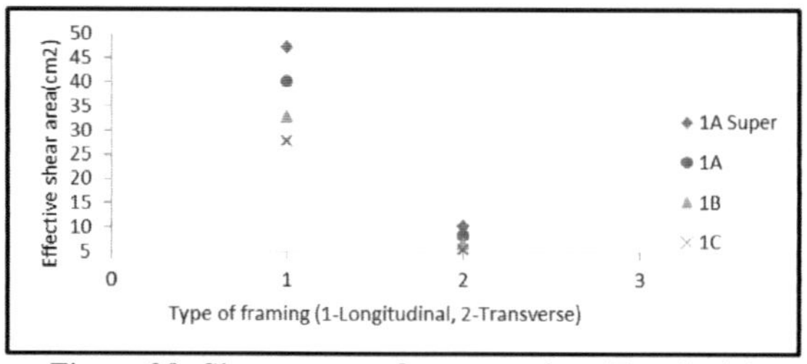

Figure 28: Shear area vs frame spacing (Fore body)

From Figures 26, 27 and 28, it is observed that longitudinal framing system gives higher ice framing effective shear area than transverse framing system which is almost 25%.

4.4 Effect of material grade on ice belt structure

Material grade plays an important role for scantling calculation of ice belt structure. Ice belt structure generally has heavier scantling. So, its of outmost importance to select the proper material grade which will reduce the lightship of the vessel. In this analysis, normal shipbuilding steels Grade A (Yield stress 235MPa), Grade AH32 (Yield stress 315 MPa) and Grade AH36 (Yield stress 355 MPa) are considered. In this analysis, in order to investigate the effect of material grade longitudinal framing system is adopted. Also, stiffener spacing and frame spacing is kept constant which is 350mm and 2.1m respectively.

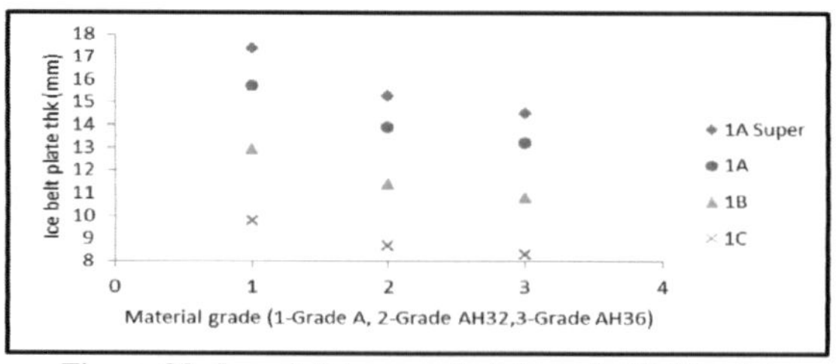

Figure 29: Ice plt thk vs material grade (Aft body)

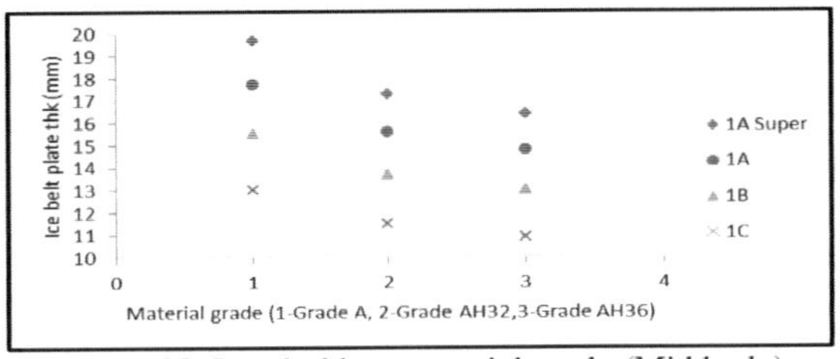

Figure 30: Ice plt thk vs material grade (Mid body)

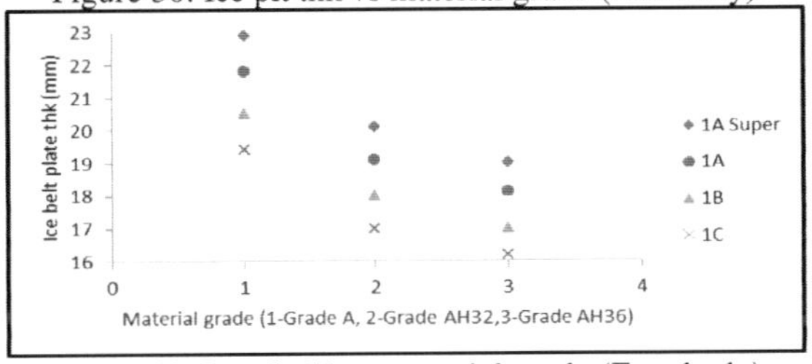

Figure 31: Ice plt thk vs material grade (Fore body)

From Figures 29, 30 and 31, it is noticed that ice plating thickness increases with the decrease of material yield strength and the increase is almost linear. Also, it is noticed that ice class 1A super needs higher scantling than 1C due to high ice load. Moreover, it is also investigated that changing material grade from A to AH36 plate thickness of ice belt reduces almost 17%.

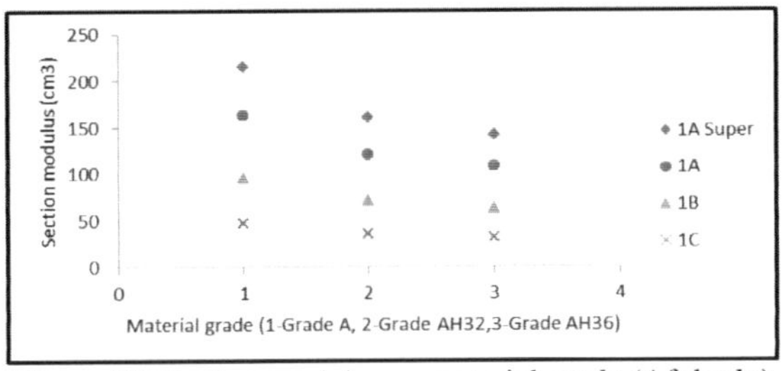

Figure 32: Section modulus vs material grade (Aft body)

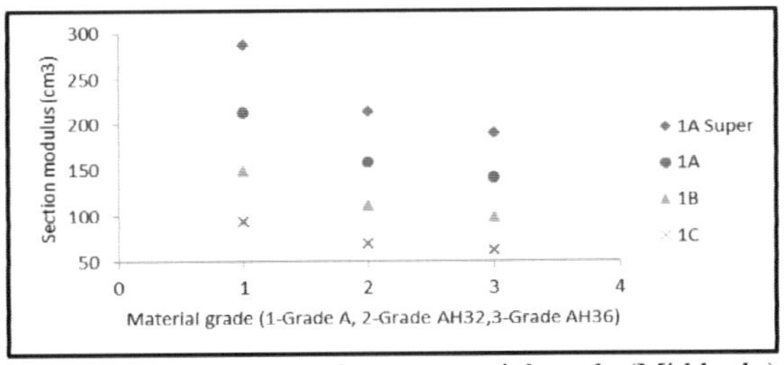

Figure 33: Section modulus vs material grade (Mid body)

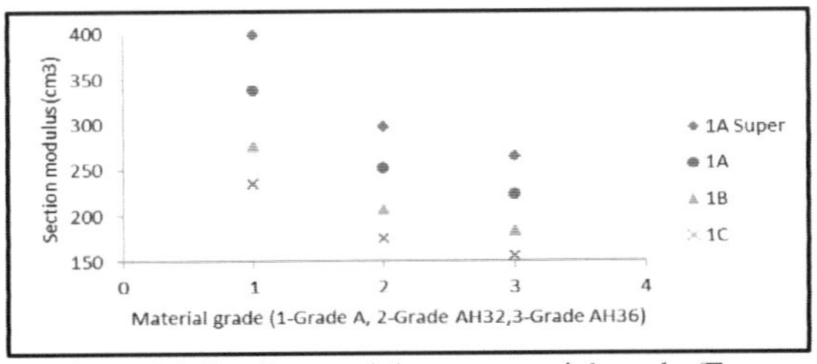

Figure 34: Section modulus vs material grade (Fore body)

From Figures 32, 33 and 34, it is found that ice framing section modulus increases with the decrease of material yield strength and the increase is almost linear. Moreover, it is also noticed that changing material grade from A to AH36 section modulus of ice framing reduces almost 30%.

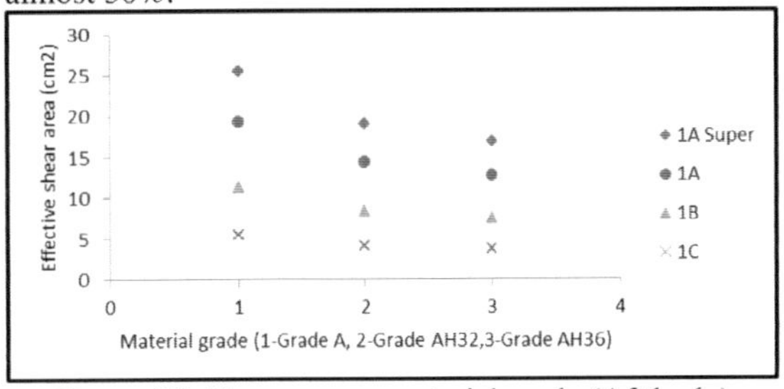

Figure 35: Shear area vs material grade (Aft body)

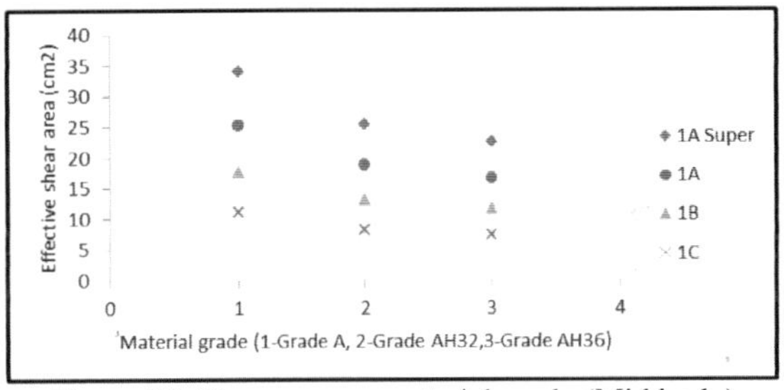

Figure 36: Shear area vs material grade (Mid body)

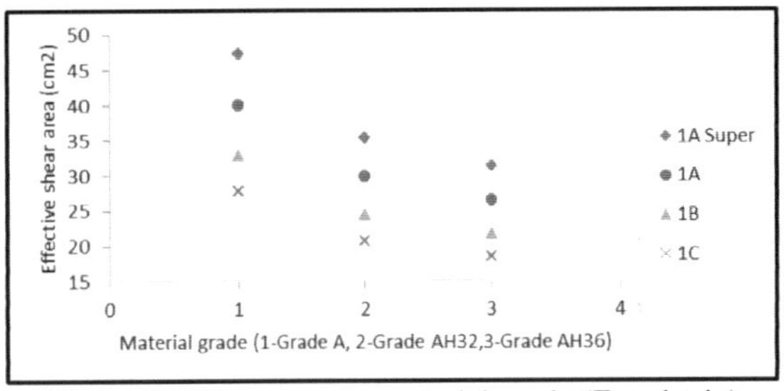

Figure 37: Shear area vs material grade (Fore body)

From Figures 35, 36 and 37, it is found that ice framing shear area increases with the decrease of material yield strength and the increase is almost linear. Moreover, it is also observed that changing material grade from A to AH36 shear area of ice framing reduces almost 30%.

5. CONCLUSIONS

From the above analysis, the following conclusions are proposed:

1. Ice belt plating thickness increases almost 10% with increasing stiffener spacing by 50mm.
2. Ice framing section modulus and effective shear area increases almost 5% with increasing stiffener spacing by 50mm.
3. Ice belt plating thickness does not change with the increase of frame spacing.

4. Ice framing section modulus and effective shear area increases almost 25% and 10% respectively with increasing frame spacing by 300mm.
5. Transverse framing system with a longitudinal main girder has an advantage over longitudinal framing system while designing ice belt structure. Longitudinal framing system gives higher ice plating thickness and ice framing section modulus than transverse framing system which is almost 20% and 25%. However, the analysis only considers the local strength required in the ice-belt area. Moreover, the weight advantages may diminish when global requirements are considered with regard to the common structural rule (CSR).
6. Ice plating thickness and ice framing section modulus increases with the decrease of material yield strength and the increase is almost linear. It is also investigated that changing material grade from A to AH36 plate thickness of ice belt reduces almost 17% and section modulus reduces almost 30%.

6. REFERENCES

[1] Filip bergbom wallin et.al, Ice Class Requirements on Side Shell Structures - A comparison of local strength class requirements regarding plastic design of ice-reinforced side shell structures, *Master of Science Thesis, Chalmers University of Technology*, 2012.
[2] K. Riska, Comparison of the scantlings derived from different ice rules in order to establish equivalencies between ice classes, *Ship Laboratory, Helsinki University of Technology, Helsinki, Finland*, 2002.
[3] Robert E. Dvorak, Engineering and Economic Implications of Ice-Classed Containerships, *Massachusetts Institute of Technology*, June 2009.
[4] R. Bridges, S. Hasolt, M.S. Kim, K. Riska, Current Hull and Machinery ice class rules requirements and impact of IACS Polar rules, *ARCOP Report W2.2.2. 16* December, 2004.
[5] C.G. Daley, A. Kendrick, E. Appolonov, Plating and framing design in the Unified Requirements for Polar Class Ships, *Published in POAC'01, Ottawa, Canada*, August, 2001.
[6] Guidelines for the application of Finnish –Swedish IceClass Rules, *TraFi, TRAFI/494131/03.04.01.00/2016, Helsinki, Finland*, 1 December 2017.

7. AUTHORS BIOGRAPHY

Sohanur Rahman holds the position of Structural Engineer at Ankerbeer Engineering and Design, Kolham, The Netherlands. He is responsible for carrying out structural calculations on ship structure. He completed his Master of Science degree on Naval Architecture from University of Liege, Belgium and Bachelor of Science on Naval Architecture from Bangladesh University of Engineering and Technology, Bangladesh. He is the author of several journal and conference papers which have been published in several Journals and conferences.

The Royal Institution of Naval Architects
8-9 Northumberland Street
London, WC2N 5DA
United Kingdom

ISBN 978-1-7138-8300-5

CURRAN ASSOCIATES INC.
proceedings.com